SEASICK

The University of Chicago Press CHICAGO AND LONDON

Seasick

OCEAN CHANGE AND THE EXTINCTION OF LIFE ON EARTH

Alanna Mitchell

The University of Chicago Press, Chicago 60637
The University of Chicago Press, Ltd., London

Paperback edition 2011
Printed in the United States of America

19 18 17 16 15 14 13 12 11   2 3 4 5

ISBN-13: 978-0-226-53258-5 (cloth)
ISBN-13: 978-0-226-53263-9 (paper)
ISBN-10: 0-226-53258-5 (cloth)
ISBN-10: 0-226-53263-1 (paper)

Library of Congress Cataloging-in-Publication Data
Mitchell, Alanna.
Seasick : ocean change and the extinction of life on Earth / Alanna Mitchell.
p. cm.
Includes bibliograpical references and index.
ISBN-13: 978-0-226-53258-5 (cloth : alk. paper)
ISBN-10: 0-226-53258-5 (cloth : alk. paper)
1. Marine pollution. 2. Ocean—Effect of human beings on.
3. Extinction (Biology). I. Title.
GC1085.M57 2009
577.7'27—dc22   2009016522

♾ The paper used in this publication meets the requirements of
ANSI/NISO Z39.48-1992 (Permanence of Paper).

This book is for James Kenneth Patterson
and his remarkable uncle Kenneth Mark Drain

Contents

Prologue

Writing this book was an act of courage. I have been afraid of water since I was a toddler and nearly drowned in an undertow on summer vacation. My parents, sitting on the beach a couple of meters from me, watched me go down and bob up a couple of times before they realized that I was in trouble. My father dashed in and pulled me out, saving my life. I don't remember the incident, but sometimes I wonder if terror that strong and that primal is stored in the depths of the brain. Certainly it pounced on me at the most surprising moments as I pondered the fate of the sea for this book, and only while I was immersed in water.

Like so many through the ages, though, just as I feared the ocean, I was beguiled by it. I was struck, not just by its mystery or its role as backdrop for the collective human imagination, but by the fact that we so rarely perceive it as a single, interconnected system, much less one that is critical to our own survival.

At first I felt like an explorer, immersed in the unknown but gradually emerging with an understanding of the sheer physicality of the ocean: how the currents move, how they shunt heat around, how the thin layer of molecules at the surface of the water breathes into a similarly slender layer at the bottom of the atmosphere, creating the chemical soup that allows life. I read with awe of the complicated architecture of the deep ocean floor, where the plates that make up the earth's surface are dying and being born, and where the planet's largest mountain range—unseen by all but a handful of humans, and unimagined by most of us—runs like a backbone through the Atlantic, the Americas on one side, Europe and Africa on the other. The global ocean was a new, secret planet hidden in a world I thought I knew well, one I had spent the better part of a decade exploring and writing about. It was intoxicating.

At the same time, the Census of Marine Life began a billion-dollar worldwide exercise to figure out what lives in the global ocean. It was a pioneering human undertaking, aided by new techniques that allow researchers to read each creature's DNA and trace its path backward to the origins of life. Month after month new ocean creatures have been discovered, new clues mined about how life as it is now came to be.

But as I learned, slowly and painfully, about how the ocean works, I also learned that we humans are changing this vast medium of life in ways that carry almost unimaginable risks for our own species and many others. We are changing the ocean's very chemistry. As the eminent marine biologist Jeremy Jackson, of Scripps Institution of Oceanography, wrote recently, we are creating a "brave new ocean" that, without profound and multifaceted changes in human behavior, will lay the groundwork for a "mass extinction in the oceans with unknown ecological and evolutionary consequences."

It would be easy to despair.

For me, it took a journey to the deeps to find a glimmer of hope. It was a scientific expedition with Harbor Branch Oceanographic Institution, a descent to 3,000 feet—the bottom of the ocean. There are parts of the ocean floor that go much deeper—to about 36,000 feet—but 3,000 feet is incredibly deep. Very few humans have ever voyaged as deep as we went that day, and as far as the Harbor Branch group knew, I was the first journalist.

The pressure at that depth is intense and would be instantly fatal if the craft sprang the smallest leak. I was terrified. But I had spent two and a half years obsessed with the ocean, this grand mother of life, and I simply had to be inside her. I needed to bear witness to as much as I could of this vast three-dimensional realm that so powerfully shapes the biology of our planet. Early-stage fetuses depend on amniotic fluid that is a chemical replica of the marine world; the composition of our blood plasma too is astonishingly similar to the composition of seawater. So I hoisted myself into the tiny aluminum chamber that would carry the ship's electronics engineer and me down to the ocean floor. We lay side by side to fit and the hatch clanged shut.

It is pitch black at 3,000 feet, and bitterly cold. I was intently focused on the composition of the air in my aluminum womb. It became a microcosm for the chemical changes to the atmosphere and ocean that this book is about, a realization of the idea that a system has limits.

I had an epiphany there. It had to do with the nature of hope. I rolled out of my watery cradle three hours later tremulous, moved, triumphant, convinced that our species, with all its failings and fumblings, is ready for its toughest journey yet.

The last best place on earth 1

GREAT BARRIER REEF

If all life on land were to vanish tomorrow, creatures in the ocean would flourish. If the opposite happened—if the ocean's life perished—then the creatures on land would die too. Life, if it went on, would have to start over.

For the vast majority of the earth's 4.5-billion-year history, there was no life outside the ocean. The planet's first creatures—bacteria-like organisms stewing in a primordial soup—appeared in the ocean more than 3 billion years ago. The first life forms on land, plants such as ferns and mosses, came into being only about 400 million years ago. Animals capable of living on land came even later.

It's hard for humans to grasp. Throughout much of our species' short time on earth—perhaps 200,000 years—we have spun stories that assure us that the world is ours to use as we see fit. Biologically, though, humans are here at the sufferance of the ocean. It gives us life, and it keeps us alive. The global ocean controls the planet's carbon cycle, determines its climate, and, through the work of phytoplankton, provides every second breath of oxygen we take.

That is not, however, how we've viewed the sea. We've tended instead to see it as the source of other things: an endless supply of food, a route to distant places. We have learned that the sea is useful, but we have not understood until the last few generations that it makes up 99 percent of the living space on earth and is also our main life-support system.

But until very recently it hasn't mattered what humans think or don't think or what stories we tell. Even what we do, or don't do, has been inconsequential in the grand scheme of things—not, perhaps, for individual humans but for the planet as a whole, and certainly for the global ocean. The planet's resilience, its vastness, has compensated for whatever we might do.

Now, though, it does matter. After millennia of insignificance, human actions and belief systems—what we do and how we think—have effected change on a scale that is damaging the ocean. Our actions are dangerous to us and to millions of other living things. We are altering not just bits of the sea with dreadful oil spills or eroding shores or vast extinctions of fish, but the whole, interconnected global system that is the ocean, the main medium of life on earth.

As goes the ocean, so goes life. This is the axiom that, in our hubris, we have refused to acknowledge. We have regarded the ocean's mysteries as elemental, immutable. Now, though, we must understand its mechanisms. Only then will we be equipped to alter our actions and stop the damage, to stand back and let the sea heal if it can. This is, in the view of many scientists, our only hope for survival.

*

One of the best places to see how the ocean should work is the Great Barrier Reef, which runs down the east coast of Australia, and that's why I'm on a lurching catamaran with a few dozen seasick tourists, braving high winds and roiling waves. To me this is the mother ecosystem, a modern proxy for where life must have begun, one piece of the ocean that is still gloriously intact. I have traveled around the world to savor it, as if it were a watery Paradise before the Fall, the last best place on earth.

Considered one of the seven natural wonders of the world, the Great Barrier Reef is by far the largest coral reef, an intertwined collection of more than 2,900 smaller reefs, including some of the oldest coral colonies on the planet. They make up a long, bony structure so substantial it can be seen from space.

It is a biological gold mine, more productive even than tropical rainforests. Within earth's most important medium for life, the Great Barrier Reef is arguably the most important part. It is the marine equivalent of the biggest city in the world, a vast maternity ward, the lushest, most productive, and probably most complex biological system on the planet. Innumerable different types of plants and animals live here. It is an engine of evolution.

How does it work?

The ocean is impossibly complicated, interconnected, turbulent, and nonlinear, and it touches every part of life. Humans can grasp its import only in terms of far simpler, smaller-scale phenomena: Every tear you cry, for example, ends up back in the ocean system. Every third molecule of carbon dioxide you exhale is absorbed into the ocean. Every second breath you take comes from the oxygen produced by plankton.

Think of a jigsaw puzzle whose pieces interlock in five or six dimensions instead of just two. Take the fierce winds today off the coast of Australia. They have kept every single tourist boat along the reef at dock, except this massive, fast catamaran. As we boarded this morning, the crew issued stern warnings about how rough the seas would be and insisted that everyone take seasickness pills—either Dramamine or ginger tablets—for the hour-long journey.

To humans these high winds are an annoyance or a danger. One woman is crying as the catamaran heaves, clutching a white paper bag in case she throws up. A young man vanishes to the rear of the ship, a slight sweat on his upper lip. Looking at the horizon is supposed to calm a seasick stomach, but for him and many other passengers on this boat, the trick is decidedly not working. Even the crew members are sighing deeply and trying to remain cheerful.

*

Throughout the ocean, the winds drive currents on the surface. Together these air and water currents circulate the sun's energy—its heat—making the planet habitable. Some of that heat moves in great, snaking submarine rivers through the ocean basins. In the north, for example, when the warm water, heavy with salt, loses enough of its heat close to the pole, it falls to the abyss, and then meanders back toward the Antarctic in a massive hemispheric conveyor belt. This is a major way that the sun's energy—the source of all life—is distributed throughout the planet.

It's a system in perpetual motion. In any bucket of water you collect at the seashore, there may be molecules that originated near the south pole—where the planet's deep, cold waters are mixed as if in a huge bowl—at the equator, or anywhere else in the vast ocean system.

Like the blood in your body, the ocean waters are constantly on the move. You have no blood that is only of the brain or only of the thumb. Nor is there seawater that is only of the Pacific or only of the Indian Ocean. There is only one ocean and it is a single system, chemically, physically, and biologically.

While the ocean has different surface temperatures at different latitudes, and therefore supports distinct groupings of creatures, each collection of life supports all the others. For example, and in simplified terms, phytoplankton (plant plankton) like the cold waters of the high latitudes. They reproduce more abundantly there. But the creatures they feed, such as larger zooplankton (animal plankton), are food for krill and other little crustaceans, and these are food for little fish, which in turn nourish the long-distance swimmers of the ocean, such as tuna, marlin, and sharks.

Moreover, at any given latitude, the surface of the ocean and its lower levels are at different temperatures and so support different creatures. Here again the winds are crucial. Powerful winds—including hurricanes—help mix water in the surface layer with water further below. Deep water contains nutrients and chemicals essential to creatures nearer the surface, particularly plankton. Without wind the water would remain stratified: nutrients would stay in the deep, and the plankton would be unable to grow. Without plankton, the food web of the ocean would collapse. All the larger creatures in the sea depend on plankton, either as their own food or as food for the creatures they eat.

Plankton may be at the bottom of the food chain in the ocean, but they are at the top of the biogeochemical system of the planet, helping to control its carbon and oxygen cycles. They are the key to how these elements move through living creatures and through their nonliving surroundings so that they can be used again by the living.

Winds also help foster the exchange that occurs when the top layer of the ocean's surface meets the bottom layer of the atmosphere. This boundary, only molecules thick, is a vast entry point where carbon and oxygen, heat and cold, and gassy water move between sea and air.

The whole system works within an astonishingly narrow set of limits and norms that have been unchanged, in some cases, for millions of years. For example, the acid-base balance of the ocean, which is intimately connected with the concentration of carbon dioxide in the atmosphere, has been about the same for some twenty million years. A pH of 7 is considered neutral. The open ocean has historically registered 8.2 on the pH scale, meaning that the water is slightly basic. At the coasts, where freshwater runs into the sea, readings sometimes hover around 7.5 before mixing with the rest of the ocean, and the assemblage of living creatures is attuned to this different, more acidic pH.

Outside of that range of acidity, however, the biological and chemical functions of the ocean don't work as well, at least not for all of the different types of creatures that are needed for the ocean system to do what it has done over time.

*

We arrive at the outer fringe of the northern Great Barrier Reef. Just beyond us, toward the open ocean, the continental shelf plunges straight down for more than a thousand meters. In parts of the frigid waters of the abyss, a new ocean floor is constantly being made by the molten mantle that seeps through the rifts.

This particular segment of the reef is about as pristine as it could be, a rare quality in the modern world's ecosystems and especially uncommon for coral reefs, which once thrived in the warm, shallow waters that circle the earth to the north and south of the equator. Turtle Bay, part of the Agincourt Reef on the outer edge of the Great Barrier, is a world of wonder. The waters are emerald green, shot through with brilliant blue. The bottom, visible among masses of corals, is white sand and only several meters below the surface.

Further south on the Great Barrier Reef, things are not quite so splendid. The worldwide decay of coral reefs—spawned by pollution from land, too much fishing, nasty practices to capture wild fish for the aquarium trade, and waters that are too hot because of global climate change—has already started to take its toll.

Over the past twenty years, living coral around the world has been destroyed five times more quickly than tropical rainforests. The pace of destruction has doubled in the past decade, and some corals in the waters surrounding the Galapagos are near extinction. More than a quarter of the world's reefs have already been killed off, and another 50 percent are in severe trouble. In the Caribbean, which was a coral wonderland just three decades ago, 80 percent of the reefs have died, reduced to limestone rubble and algal slime.

This level of destruction is unprecedented in modern times. We have to look far back in the fossil record to see a similar coral death rate. In some earlier cases of wholesale die-off, it took millions of years for the coral to come back, seeded from handfuls of hardy remnants that managed to survive until conditions improved again.

The boat's crew is talking up the lazy turtles we can expect to see, the Volkswagen-size clams that take two hours to shut their mouths, the shoals of brilliantly colored tropical fish. For me, though, it's all about the corals, those ancient, primitive animals whose ancestors have been on the planet for about 450 million years.

I zip myself into my diving skin, shove my feet into the flippers I've carried halfway around the world, spit into my mask to prevent it from fogging up, and plunge under the fierce waves.

What I see is stunning. And sheer calm after the messy voyage to get here. Katharina Fabricius, a coral scientist from the Australian Institute of Marine Science in Townsville—and my tutor and host on this trip—has given me a primer on what to look for. A few days ago at her house on Magnetic Island, she ran me through her "good reef/bad reef" presentation, a series of underwater photographs she's taken and used, since 2001, to explain to politicians and others some of the problems with today's reefs.

A good reef has lots of structure, she says, high coral cover, lots of different types of coral, lots of baby corals, and lots of other marine organisms seeking shelter. A bad reef has little structure. It's flatter, with fewer types of corals, some dead corals, empty patches, few fish, and baby corals smothered in mud. Go to Agincourt Reef, she advises me. It's still in great shape.

She's right: Turtle Bay is an excellent reef. Every square centimeter is full of corals, masses of different shapes, colors, and sizes. There are no patchy, thin, crumbling spots as there are on so many other corals reefs, including those I have seen in the Caribbean and Asia. No dead corals, and no suffocating slick of big, indigestible algae.

Here are dozens of types of hard coral, each of which has built up a substantial bony structure using the calcium and carbon stored in the ocean. These limestone shapes will keep growing for as long as the coral animal—a thin membrane of flesh attached to the surface of the limestone—stays alive.

Some faster-growing branching corals make shapes like the limbs of trees. I see field upon field of *Acropora* species, like the antlers of elk and deer, some of which are neon blue. I've seen beds of *Acropora* in the Caribbean that were nothing but skeletons rotting in the saltwater. This richness is breathtaking by contrast.

Other of the corals here are the slow-growing, massive types that build the foundation of the reef. They have nooks and slits, caves and secret crevices running along their meters-high structure—each filled with other species, a riot of color. They are like underwater coral gardens, some so close to the surface that I have to be careful not to kick them with my flippers as I snorkel over the top.

These bony shapes are necessary if a reef is to be a home and protector and on-demand food market to the sea's creatures. Once you've got the structure, Fabricius says, the soft corals come, and the fish and the worms, the mammals and bacteria and reptiles, the birds above and all the other creatures that lay their eggs and rear their young and use the reef to survive. All told, about a quarter of the sea creatures that humans catch commercially spend some part of their life cycle on a coral reef.

The Great Barrier Reef is home to so many plants and animals that a full count has never been made, not even of the corals themselves. Many are still unknown to science and unnamed. Yet, while the reef is critical to the survival of so many other species, it is also totally dependent on plants for its own survival. The coral animals that build the reef cannot exist without the brightly colored zooxanthellae that live within their very cells. Like other plants, these microscopic algae convert energy from the sun into food through a chain of chemical reactions. The food they produce feeds the coral animals as well as the algae. Without the algae, the corals starve and lose their brilliant color. Without the corals for protection, the algae die. It is the canonical example of a symbiotic relationship.

How well could the planet's other creatures get along without corals? What will happen to the physical, chemical, and biological structures of the ocean if the reefs die? The answers are unknown.

As I snorkel here, marveling at the virtuoso display of biological creativity, I wonder whether humans are symbionts who have lost the knowledge that we depend on other creatures for the basics of life support. In terms of pure self-interest, this is a problem. If we depend on corals, algae, plankton, and millions of other species, and if we are killing them off, how will we survive?

The textbook definition of a parasite is this: a species that thrives while its host dies. Unfortunately, the host in humans' case is not a single creature but all sorts of species and even whole ecosystems. In biological terms killing the host, or the host system, is a risky strategy. That is, unless there are other systems or hosts for the parasite to move on to.

I wonder whether these simple, ancient animals, which do not even have brains, can remind humans about the ground rules of symbiosis and the perils of ignoring them. What if the Great Barrier Reef is not just a vision of the unspoiled past but a cautionary lesson, foreshadowing future destruction?

*

It was snowing when I left Toronto. But on Magnetic Island the temperature is sweltering, in the 30s Celsius. It is Easter weekend, the busiest of the year on an island renowned for beaches, surfing, and honeymoon getaways. Every hotel room is packed and the beaches are thronged with holiday makers. Katharina Fabricius, a slight, athletic woman with the fine features of a runway model, has picked me up after the long trans-Pacific flight and offered me a bed on the floor of her spare room.

Fabricius's house is in a small thicket. She and her partner, Glenn De'ath—also an eminent ocean scientist—dislike doors and windows and walls. Geckos, spiders, possums, and the occasional snake wander in and out at will. Breezes sweep through the house. In a high corner of the porch where they eat their meals, the couple has built a shelter for tree frogs so they won't be picked off by predators. The night I arrived, awakened by thumps on the floor, drunk with jet lag, I bolted from my room with a wavery flashlight to see that a scatter of the fat, neon green frogs had jumped from the ceiling onto the floor of the open kitchen.

Here in her home, Fabricius explains some of what this great reef can teach humans. The reef is one of the places where all of the pressing threats to the ocean intersect, a microcosm of problems and also of hopes. The plain fact, she says, is that the Great Barrier Reef is in mortal danger.

It's not just the obvious, localized threats that have affected so many other reefs around the world. The Great Barrier Reef has been designated an Australian national park and is protected by rigorously enforced laws against many time-honored coral-killing practices: catching too many fish, trawling, long-line fishing, and the use of cyanide poisoning and dynamite to catch tropical fish for the marine aquarium trade. Fully a third of the reef is a no-take zone, meaning that no commercial activity is permitted and that other activities are controlled. No, the biggest threats will come from what Australian law alone cannot prevent: global climate change.

Climate change has three direct effects on corals that scientists have become aware of in recent years.

First, the high concentration of carbon dioxide in the atmosphere warms the ocean to the point that the corals' symbiotic algae die. That's known as coral bleaching, because when the algae die, their color disappears and leaves the corals' flesh ghostly white. Unless the corals can attract new algae they starve, and the flesh covering the bony skeleton rips off in shreds. Devoid of animal cover, the skeleton itself then begins to decay. Even if they can attract new microalgae, the bleached corals' ability to grow and reproduce is impaired while they heal.

Coral bleaching—recognized by scientists only a couple of decades ago—has been increasing all over the world as climate change takes greater hold. On the Great Barrier Reef, waters have warmed on average less than half a degree Celsius since the end of the nineteenth century, yet bleaching has run far ahead of expectations. By the end of this century, ocean temperatures at the reef are forecast to increase another one to three degrees Celsius, leading to a dramatic increase in bleaching. The best predictions say that by 2050, severe bleaching will be a yearly event on the reef. With no cooler intermediate years, the corals will have no opportunity to heal and grow back or to restore their ability to reproduce.

Second, climate change threatens to increase the volume of the sea. As land-based ice sheets in Antarctica and Greenland melt in the warmer climate, extra water will pour into the global ocean basin. It's not clear how much higher sea levels will rise or how quickly—estimates vary from far less than a meter to as much as 9 meters in the coming century—but odds are that the world's remaining healthy reefs will be deeper underwater. That will reduce the amount of sunlight that reaches the algae, which may prevent them from carrying out photosynthesis; as with direct warming, the algae will die. In effect, the corals will drown.

Third, and even more serious, are the chemical changes occurring in the ocean water itself. Scientists are calling this the gorilla in the corner, the problem that eclipses even global climate change as a threat to the planet's current forms of life. Atmosphere and ocean are ineluctably joined. In fact, some scientists consider them parts of the same system. They interact with each other, exchanging chemical compounds much as oxygen and carbon dioxide are transferred between air and blood within the body. When we burn fossilized carbon in the form of coal, oil, and gas, carbon dioxide gets stored both in the atmosphere and in the ocean. The ocean has absorbed about a third of the extra carbon dioxide that we have pumped into the atmosphere since we started using fossil fuels, digging up grasslands, and cutting down forests.

Today, the concentration of carbon dioxide in the atmosphere is about 387 parts per million by volume. Before the industrial era it was 280 parts per million. The latest intergovernmental panel report on climate change has predicted that once the level reaches 450, possibly by midcentury, humans will have pushed roughly a quarter of the planet's creatures into extinction. If it reaches 550 parts per million—about double preindustrial levels—we will have caused the genetic extermination of up to 70 percent of living things. Such a massive spasm of extinctions would not be without precedent, but we know of just five others in the planet's 4.5-billion-year history.

In the atmosphere carbon dioxide is chemically inert, but it has a huge effect on climate. In the ocean carbon dioxide is chemically active; it combines with other compounds in saltwater to make the ocean more acidic. As more carbon dioxide enters the ocean, the water's acidity increases. The changes so far are relatively small—about a tenth of a unit of pH on average across the global ocean. That sounds tiny, but the pH scale is not straightforward. In mathematical terms it is logarithmic rather than linear. This means that small changes in the measurement correspond to enormous changes in the actual chemistry: a pH of 6, for example, is 10 times as acidic as a pH of 7, and a decrease of just a tenth of a unit represents a 30 percent increase in acidity. Depending on how we continue in our use of fossil fuels, the pH will have dropped by four- or five-tenths of a unit by the end of this century—a historically massive amount.

Because of the way carbon and calcium interact in the ocean waters to form new compounds, calcium will be less available to creatures that need it to make bones and skeletal structures—like corals. That means corals will not be able to make reefs as quickly. Eventually, they may not be able to build their structures at all. Once the pH of the ocean reaches a certain point, the water will become acidic enough to corrode the healthy reefs that remain. Like the ancient pearl Cleopatra famously dropped in a goblet of vinegar so as to dazzle Antony with her decadent excess, the corals will dissolve.

Of course, once the corals are gone, the creatures that depend on them will be gone too.

Corals thus face multiple threats. These will not be fully realized for years or even decades. But they are real, and they tend to magnify each other's negative effects, creating the potential for a spiral of destruction on the world's reefs.

A medical metaphor may be helpful: When the human body has a bacterial infection, it develops a fever to protect itself. The high temperature makes it hard for the bacteria to survive. The goal of the fever is to kill the infectious agents and restore the body to health. But if the infection rages too far and the fever goes too high, what's known as a positive feedback system sets in and the body switches from fostering health to courting death. Rather than preserving the old order, it gives birth to a new and different system—decay—which is designed to break the body into its component chemicals and offer them up for reuse by new forms of life. Fever takes over and the body dies. The old system is gone forever and cannot be recovered.

In the case of the Great Barrier Reef, the system is striving to maintain its current health. When its corals bleach horribly, as they did in 1998 and 2002, the reef sets about healing itself, preparing to grow again. But if the predicted miseries occur—increased acidity, bleaching, drowning, more intense cyclones and the systemic disturbances that will accompany them—then the reef will move into a new mode. Instead of trying to preserve the regime that exists now, it will begin to help a different one come into being.

In the Caribbean, where 80 percent of corals are dead, this is already happening. A new system has taken over in which slimy algae are the main form of life, and corals, fish, and mammals are relatively rare.

Scientists believe that the corals won't be able to adapt to the onslaught of carbon dioxide if humans fail to curb emissions. Ocean chemistry, for example, has begun to change dramatically from decade to decade, far more quickly than it has for millions of years. Even if corals could evolve quickly, the new environment is a moving target. But most corals grow slowly, reproduce with difficulty, evolve in minuscule increments. And they appear not to have adapted quickly to changes in ocean chemistry in earlier periods of great disruption. The fossil record shows that when corals vanished they did so for millions of years at a time, coming back only because a few species held on somewhere, adapted painfully slowly, and eventually formed new populations.

To halt the coming damage to the reef and the multitude of creatures that depend on it, humans would have to ensure that the concentration of carbon dioxide in the atmosphere rises little higher than it is now. That would mean switching to sources of energy not based on carbon, or using current sources in ways that do not emit carbon dioxide into the atmosphere. And that would require rapid changes in our belief systems. We would have to acknowledge that our population and our use of fossil energy cannot keep growing, and we would have to act accordingly.

All of this is possible, Fabricius says. In fact, she argues that we have an ethical and moral responsibility to do what we can to prevent the death of the reef, with all its consequences. Then she points across the water to the Australian mainland. Further inland are coal mines that send boatloads of the carbon dioxide–spewing fuel to China. Australians glory in their powerful economy, she says, but the economy is strong partly because it provides a climate- and ocean-altering fossil fuel to another country.

Tradition, she believes, will trump reason on this issue. The climate and ocean system are changing in ways they have not for millions of years. But we will cling to our belief in growth, born hundreds of years ago, with catastrophic results. We haven't learned how to take the long view.

*

Back at Turtle Bay we're having a rough time snorkeling. The winds have whipped the heavy waves into a frenzy. Saltwater splashes through the breathing tube into the mouthpiece, forcing us to surface quickly and frequently, and gulp desperately for air. Our umbilical cords to the surface are treacherous.

I know that our own blood plasma is chemically a dead ringer for seawater—a legacy of the fact that all creatures, including our ancestors, evolved in the ocean—but as I spit out another mouthful of salty water I am forcibly reminded that this is no longer our medium of life. It is a foreign and frightening place where humans have few defenses.

We're being bounced along by the energy of the waves too. I have to learn to let myself go, to try not to fight against the water or control it. I cannot bend it to my will.

Being here requires humility.

I've just spotted a green turtle swimming smoothly among the gorgeous swaths of coral. It's fully a meter across, as big as a dining room table. I swim beside it for a while, envious of its easy movement, the graceful paddle of its feet. I'm careful not to swim above it. That can spook turtles, and sometimes they panic, dive deep, and suffocate before they can get back to the surface.

Like so many types of sea turtle, the greens are globally endangered, often killed by long-line fishing techniques and by people who crave their meat and eggs. One of its common names in French is *tortue comestible*, "the turtle that's good to eat." Only about 200,000 nesting females are still alive anywhere in the world.

Over there is a long, ribbon-thin, neon yellow fish with a black eye and a mouth like a flute. Here, a clutch of giant clams, some with intensely blue lips. And everywhere, schools of fish in colors that bring to mind candy shops and party frocks.

We're out of the water now. We've had to abort a snorkel excursion further down Agincourt Reef at a fish hot spot known as Castle Rock. By now, at the end of the day, the waves are just too heavy, the wind too strong.

I've peeled out of my diving suit and changed into dry pants and a shirt. Despite the heat of the day I am shivering in my warm jacket. Being in the ocean for hours on end gives a chill to the bone. My skin smells like salt and fish. Little abrasions I had before I went in are healed up as if they never existed.

A photographer who dived with us today and snapped our pictures underwater is flogging DVDs with pictures from Agincourt Reef that he and others took before this trip. I notice that they're mainly of fish. The corals could be so much wallpaper.

This tourist outfit, Quicksilver Group, is superb. It is renowned in the scientific community for its help with marine research. Its literature boasts that it has Australia's largest team of marine biologists outside of the government agencies. Yet there has been not a single mention of the threats the reef faces, apart from adjurations not to touch it or cart any of it away because of the hefty fines, as if tourists were the greatest danger.

I think about the meeting I had yesterday with Paul Marshall and Johanna Johnson in Townsville. They work for the Great Barrier Reef Marine Park Authority, assessing the effects of climate change on the reef, part of a larger Australian team researching this topic. They have international reputations in their field. And they are frank about the fact that they are buying time for the reef. That means limiting fishing, restricting shipping, and minimizing other local forms of damage to the reef. The theory is that if the stresses that can be controlled are kept to a minimum, the reef has a better chance to survive the massive stresses still to come as the climate continues to change.

If any reef in the world has a chance to survive what's coming, it's the Great Barrier, and these scientists and their park authority are going to give it every possible chance. Marshall lists all the natural advantages the Great Barrier Reef enjoys, compared to other reefs:

Its sheer size means that fragments of it might survive even after substantial destruction.

Water circulation—and therefore the free movement of creatures—is high around the reef. There's always the chance that if nothing survives here, a new reef could be seeded from something that survived elsewhere.

Relatively few people live close to the reef. The amount of chemical pollution and silt from industry and farming is lower than it would be if Australia were as thickly populated near its coast as, say, Indonesia or Florida.

And the reef is fabulously rich in species, seeded from the reefs of Papua New Guinea further north. Some coral reefs start with not nearly as much. The more species there are, the greater the likelihood that some of them will survive.

One of the terms that keeps coming up in our conversation is *refugia*. It refers to little clusters of genetic coral material that will slip through the cracks of destruction and live to build another reef. Marshall and Johnson are counting on a few little bits of reef being left somewhere.

Still, they know that the Great Barrier Reef is in for some severe changes in the coming years. It's locked in, as they put it. Already Australia's monsoon season has begun to fail. Clouds are vanishing and the temperature of the air, sand, and water is rising. There's also less wind to cool things down.

They already see changes in the system reflected in what's happening with seabirds and sea turtles. In 2005, the hottest year on record on the Great Barrier, baby birds died en masse. At first the scientists thought that the chicks had died of exposure in the torrid heat. But the problem turned out to be something quite different. The waters on the reef were so hot that the plankton moved to cooler places and the fish followed. So when mother seabirds went foraging for food to feed their chicks, the little fish weren't there. The chicks died of starvation.

For the sea turtles 2005 held a different but equally grisly fate. When scientists couldn't figure out where all the baby turtles were at Mon Repos beach, the largest mainland turtle rookery in Australia, they began searching the nests. They found cooked eggs. It had been so hot that the eggs baked instead of hatching.

Marshall sees hope, though. Scientists have a huge role to play in crafting programs to remove what stresses they can and in pooling knowledge to find solutions for the problems of the reef. He likens it to a Darwinian age, an era of scientific discovery. "Even if we can make the difference between 5 percent coral cover and 15 percent, it's well worth our while," he says.

I look around me at these reef aficionados on the catamaran, munching on their sweets and cheese. What would they do if they knew all this? Would they be shocked? Would they make a ruckus, push for solutions?

I have the absurd urge to get up and shout, "Does anybody on this catamaran know what's going on here or why it matters?" Instead, I vow to team up with a group of scientists who will be examining parts of the Gulf of Mexico that have lost their oxygen, become devoid of life. I will voyage from this Eden to a dead zone. My hope is to garner more clues about how this complex ecosystem works.

Afternoon tea is ending. The crew is again handing out doses of Dramamine and ginger for the rough trip back to the mainland. We will speed bravely, boldly, through the stiff waves over the corals, the turtles, fish, sharks, and giant clams. It will be as though we were never here.

Already, many of the passengers are dozing.

Reading the vital signs: Oxygen 2

GULF OF MEXICO — Life and death in the blob

Not far from New Orleans, where the mighty length of the Mississippi River discharges like a fire hose into the Gulf of Mexico, lies the blob, a thick, dense, shape-shifting layer of water stuck beneath the surface. Parts of the blob have no oxygen at all, an astounding phenomenon for such an energetic part of the global ocean.

Conceptually, the blob is like those lava lamps from the 1960s. When the lamp was plugged in, heat from a lightbulb melted wax, which then rose and fell within a cone of colored liquid, forming funky three-dimensional shapes. The blob is like the wax, a discrete, immiscible mass that changes its size and shape each year. When I visit, it's settled thickly on 17,000 square kilometers of the gulf's seafloor but rises in places nearly to the surface.

Fishermen call the blob a dead zone. When they trawl the bottom for fish, shrimps, and crabs, they get nothing. But it's not clear whether the entire water column is dead; only parts of it are wholly lacking in oxygen. The questions my hosts are investigating are what exactly can live here, with little or no oxygen, and how the extent of the dead zone affects the creatures that ought to be present. Do they suffocate, move to the edges of the blob, where there is at least some oxygen, or swim in and out of the oxygen-deprived zone? Does the normal collection of creatures change for good? Does the blob affect the overall capacity of the fabulously rich Gulf of Mexico to support life? What are its implications?

I'm with a group of American scientists who have allowed me to join them on an expedition to examine the blob in detail for the first time. We'll be sampling the water both within and outside its boundaries, to measure the difference a lack of oxygen makes to what can live, not just within the deoxygenated area, but throughout the water column. It will take us eleven days of around-the-clock work at sea to gather data from a small, meticulously mapped portion of the blob.

There are about 407 dead zones in the world, a figure that has doubled each decade since 1960. This one, in the Gulf of Mexico, off the southern coasts of Louisiana and Texas, is the second largest in the world and among the most extensively studied. There are still many, many questions about how it and other dead zones work, but building on what is already known has become increasingly imperative. The spread of dead zones throughout the global ocean is in part a result of chemicals we put into the water. But recently several new dead zones have emerged that, to the astonishment of scientists, are directly related to global climate change.

One big surprise has been a low-oxygen zone discovered at the northern edge of the California current system off the western coast of the United States. This huge, rich part of the global ocean—one of just four major eastern boundary current systems in the world—had not previously been known to have areas of low oxygen.

All four such systems are driven by the winds. When the winds whip up the currents, the water brings key nutrients, including nitrogen and phosphorus from the deep ocean, to the sunny surface to feed plankton and other creatures. When the wind patterns change, a direct result of the climate change that is warming the ocean, the circulation of water, and consequently of nutrients, can change too. That's what's happened off the northwest coast of the United States starting in 2001. A low-oxygen zone spread through the water column there and stayed for months at a time, leaving the area devoid of fish and causing a mass die-off of crustaceans.

A similar phenomenon is occurring in two of the other eastern boundary current systems, the Benguela, in the south Atlantic off Africa, and the Humboldt, in the southeast Pacific (the fourth is the Canary, off the northwest coast of Africa). The implications are frightening, because they point to the possibility that low-oxygen zones will spread rapidly, abruptly changing parts of the ocean critical to the marine food web.

Scientists are also finding that vast oxygen-starved zones in the deep waters of the Pacific and Atlantic oceans are thickening and moving nearer to the surface. That in turn keeps the colder, even deeper waters—which contain all the food and nutrients necessary for plankton—segregated from the surface. These are the unpredictable systemic ocean changes that Tim Flannery, an Australian biologist, warned me about when I talked to him about this book, the potential triggers for a large-scale switch of ocean systems. The situation is reminiscent of the Permian extinction of 250 million years ago, the "Great Dying," when ocean life nearly died out.

For me, the dead zone is another of the places where all the themes of the ocean switch converge. It is a chemistry problem, but even more it is the story of what happens when humans push a living system past its limits to the point that its chemistry goes wrong and a new system is initiated.

*

The iconic Mississippi River, which divides the United States roughly in half, has the third largest drainage basin the world, after the Amazon and Congo rivers. Each spring, as farmers sow their crops in fields that line the river's tributaries, they top them up with fertilizers: synthetic nitrogen compounds, as well as phosphorus. Over the past five or six decades, as the American population has grown, this load of agricultural chemicals has increased by a factor of at least three. Those chemicals, far in excess of the amount nature creates on its own, run off the land into the tributaries and collect in the Mississippi, which sweeps them down to its mouth near New Orleans and deposits them in the Gulf of Mexico.

The plankton and bacteria take it from there. Nitrogen and phosphorus are superfood for plankton—the ocean's tiny plants—just as they are for the plants the farmers are growing as crops upriver. The plankton gobble up the nutrients, reproduce like crazy, then die and fall to the bottom. There, on the Gulf's continental shelf, the dead plankton become a feast for bacteria, which go on an eating and decomposition rampage. They multiply exponentially and use up all the oxygen. In this region, the ability of the ocean to handle all that extra food has failed. It has reached its limit.

Under most circumstances this wouldn't be a problem; a fresh supply of water and oxygen from the ever-flowing ocean would rush in and restore the balance. But here in the vast Gulf of Mexico, the microorganistic gluttony happens in layers of water that stubbornly resist mixing. It often takes a hurricane or other storm to churn them up and put oxygen back into the water column. The blob does dissipate over the winter, once farmers stop adding nitrogen and phosphorus to their lands, but it reemerges each spring.

In some of the more than 400-plus dead zones, the nitrogen feast comes from industry, human waste, or chemicals from the burning of fossil fuels that settle in river basins, rather than from the 120 million tons of agricultural fertilizer used each year around the world. In others, the depletion of oxygen results from climate change. Dead zones in these once-productive coastal seas were all but unknown until after World War II, but now the assaults have become too much for the ocean to handle. The United Nations Environment Programme, which tracks dead zones, has warned that their number is poised to escalate rapidly and that they will greatly endanger humanity's ability to feed on fish in this century.

The Gulf of Mexico is one of the United States' most productive fishing areas, with nearly a billion dollars' worth of fish and shellfish caught each year, about a fifth of all the marine creatures caught in U.S. waters. Government officials—who finance most of the research—are keen to make sure that the dead zone doesn't interfere too much with the fisheries.

*

Stu Ludsin, a fisheries biologist at Ohio State University's Aquatic Ecology Laboratory who previously worked for the U.S. government at the Great Lakes Environmental Research Laboratory in Ann Arbor, Michigan, is the chief scientist for the expedition. It's his first time in the role. He will be conducting his own research as well as plotting the course for the trip, consulting with the ship's crew, and mediating any disputes among the scientists' big personalities.

Tonight, here at Cocodrie, Louisiana, a stone's throw from the Texas border, we are loading the *Pelican*, the research vessel that will be our home for the next eleven days, with truckloads of underwater measuring and trawling devices, computer equipment, a chest freezer, and lab paraphernalia. This activity involves winches, hard hats, and immense amounts of sweat in this drenched bayou night.

Ludsin recently competed in a triathlon in New York City that included swimming in the crazily polluted Hudson River, and he's still talking about the sludge he had to extract from his ears and nose. But he's keen not to lose his training edge. So from the van and trailer that he has driven from Michigan, he pulls out his bicycle, with a stand that will turn it into a stationary rider, hoists it onto his shoulder, and carries it up to the chief scientist's stateroom. William Boicourt and Mike Roman, two senior scientists from Horn Point Laboratory at the University of Maryland Center for Environmental Science, are familiar with the exigencies of being chief scientist on such a complex expedition. They nudge each other, bemused by the optimism Ludsin's action betrays, but remain mute.

There are three teams of scientists on this trip. Frank Jochem from Florida International University (he's since moved to private industry to make biofuels from algae) and Peter Lavrentyev from the University of Akron in

Ohio will be tracking the fate of microplankton in the dead zone. Tonight they are reverently unpacking a new instrument for its first run. Using a laser beam and a flow of ocean water piped in from outside the ship, the FlowCam will count, examine, photograph, and sort some of the tiniest creatures in the ocean. Until recently, Jochem tells me, scientists couldn't see these tiny plankton alive in the wild. They were just brownish blobs, dead on a microscope slide. Now, with this beauty he's unwrapping from its factory packaging, he'll be able to see their color and shape, along with a raft of other characteristics. This will allow him to link what's happening on a microbial level in the dead zone with what's happening among the larger creatures.

In another air-conditioned lab, a couple of meters away on the other side of the *Pelican*, Roman and Boicourt are duct-taping laptops to the counters. These will read the information on the larger algae and zooplankton that the pair and their team will collect using an instrument that flies in smooth waves from near the bottom to near the top of the water column. It's called ScanFish, and Boicourt, who has babied it through fieldwork, breakdown, and crisis for fifteen years, is anxious to make sure it's in top shape.

Ludsin and his team of grad students will be trawling for fish both in the middle of the water column and at the bottom. In contrast to the high-tech enterprises of the other teams they're in for a lot of heavy lifting and fish guts.

It's late. Everyone is exhausted. We pile into our bunks and sleep if we can as the *Pelican*, clanging with unfamiliar noises, makes its way to the dead zone.

*

On the first full day at sea the focus is on the figure of 2 milligrams of oxygen per liter. That measure is the dividing line between life and death in the ocean. Below that concentration animals can't live. Above it, they can. We are right in the center of the dead zone, so the oxygen level is uniformly below the critical figure.

Roman and I will be a team for the rest of trip, sharing two four-hour shifts a day, mainly monitoring ScanFish. He's teaching me the ropes. ScanFish measures depth, salinity, temperature, oxygen, conductivity, time, and fluorescence. The last is a proxy for the concentration of plankton in the water. Later Roman, Boicourt, and their team will analyze their findings to figure out what's living in the water column and how it relates to the oxygen levels.

Part of scientific method is to find out something and then see if you can find out the same thing another way. For that reason, this floating laboratory will use a variety of means to examine the dead zone at the same time but from different perspectives. In addition to checking the water's chemical properties using ScanFish and other instruments, looking at plankton through the FlowCam, examining algae and zooplankton (again using ScanFish), and physically collecting samples of water and fish, we will be measuring the prevalence of fish in the water column using acoustics. The acoustic instrument stays in the water almost all the time, scattering sound waves that beam down and bounce back up. The returned waves are then translated into colors on a screen that give a picture of life in this part of the ocean. If the screen is red, that means there are lots of living organisms. Bits of green indicate that there's something living in the water column. If the screen is blue, there's nothing. Right now there's a thick layer of blue on the bottom. It's spooky: this part of the ocean ought to be teeming with life.

Ludsin is on deck in the torrid heat getting ready for his inaugural trawl. He has a vast net fitted with a set of heavy doors, to keep the net's mouth wide open, and chains, to take it to the bottom. It's a cumbersome and dangerous rig that could easily crush a foot. Clad in hard hats and neon orange life jackets, he and several men winch up the net and maneuver it tail-first into the water behind the *Pelican*.

When the net comes back up fifteen minutes later, Ludsin is waiting there, wearing Wellington boots and garden gloves to protect himself from jellyfish stings. He has a red and white cooler at the ready for any fish he wants to preserve for further study. He hoists the trawl onto the deck, ready to sort through the catch.

In a healthy part of the gulf there would be hundreds of kilograms of assorted fish, mainly small ones that school in tight clusters near the bottom. He has several jellyfish, which do well in low oxygen, and an Atlantic bumper fish, maybe 2 centimeters across, that got caught in the mesh. Roman peers at it. "Put 'im on a cracker and I'll eat 'im," he says.

Ludsin, who hasn't slept in thirty hours, is discouraged. Why are there no fish at all—not even those that could live above the dead zone? It's clear that the lack of oxygen has made the fish change their behavior, but it's not at all clear what that will mean to the fish over time. How much of their life's energy are they investing in leaving this part of the sea and finding somewhere more hospitable?

Meanwhile, Roman and Boicourt are launching ScanFish. Yellow on top, orange on its belly, with fins of black, it shoots 256 light beams into the water to see what's where. Its front bears a flame decal that Boicourt's staff pasted on a few years ago when he turned fifty-six. It's not Boicourt's shift, but he's here anyway, making sure that all is well. He bows down slightly before ScanFish goes into the water, touching its flank.

Any underwater trip is fraught with risk. ScanFish is built to sense obstructions on the bottom and veer away from them, but there's so much here to hit. The seabed is strewn with wrecks and pieces of oil platforms damaged in hurricanes, not to mention thousands of buried pipelines. As well, I can see from a government navigational chart posted in the galley that the gulf is littered with unexploded ordnance and dumping grounds for explosives.

The instrument goes in. We hold our breath.

All is well. It flies gently through the water, swooping from top to bottom in elegant sinusoidal waves, collecting data.

Over on the other side of the *Pelican*, it's a different story. Jochem and Lavrentyev have been up all night trying to get their new FlowCam to work. No luck. They finally went to sleep in the wee hours without having resolved the problems. In scientific terms, this is high tragedy. Jochem, so buoyant last night as he was setting up, is barely monosyllabic now. He's hunched over the recalcitrant computer, his brow black and furrowed.

Ludsin is trawling again. He's checked the nets thoroughly. The speed is perfect, the sea as calm as it's possible to be; this system is working. He should, he figures, be catching some fish. He doesn't want millions of fish, he tells me, just enough so that he can examine them later to see whether they're getting enough to eat. Finding out requires that the creatures die, but he hates to kill them and throws back as many as he can.

He throws the trawl in. Fifteen minutes later, he and his team are heaving in the nets. It's still 29 degrees Celsius and their faces gleam with sweat. And then . . . nothing. The nets are empty. "There's no fricking fish," he mutters.

It's near midnight, the end of my shift. Jochem has finally got some pictures, even if they're not the clear, colored ones he was looking for. I wander past him and out onto the deck. The hot air packs the wallop of a fist after our air-conditioned lab.

We've been on hurricane watch today. A big one, Chris, has just broken up over Cuba, and we're waiting to see whether it will pick up some of the unnaturally hot water from the gulf and gather speed again. It's near the first anniversary of Hurricane Katrina, and this whole part of the world is on high alert.

Surrounding the *Pelican* is a city of oil derricks, shining in the humid night for kilometers around, pumping oil and gas from beneath the seafloor. The Gulf of Mexico accounts for roughly a third of the oil and a fifth of the gas produced in the United States. Some of those petroleum products go into the fertilizers that farmers up the Mississippi use to grow their crops and that in turn feed this dead zone. Some are used for fuel. All of these uses add to the greenhouse gases that are contributing to global climate change with its warmer oceans, expanding dead zones, and more intense hurricanes.

One of Ludsin's graduate students comes out on deck and looks over the side of the ship. He points. There, in the ship's light, we can see a couple of dozen small fish swimming alongside. We gasp with delight—this is the first tangible sign we've had that there is life in this part of the ocean.

*

The next day there's a squall, the residue of Hurricane Chris. We pull in all the instruments and wait it out. The sky is purplish black, and the clouds are leaden.

Some of the team have been battling seasickness since we set out and this weather makes it worse. It's a major topic of discussion: ginger versus Dramamine versus wrist zappers. Everyone is trading war stories about the worst stomachs ever and the longest recovery time. It's not easy for most human beings to be on the open ocean. One of Boicourt's lab technicians has been green and horizontal since she arrived. Today will be a rough day for her.

By nighttime we've moved on and so has the worst of the gale. It's still raining torrentially and there's lightning and thunder, but we're in water that has oxygen now, just above the dividing line of 2 milligrams per liter.

At 11 p.m. Ludsin loads up the trawl and sinks it to the bottom. Water is pouring off his hard hat and yellow slicker, and the deck is treacherous. Earlier this year, another ship he was aboard was hit by lightning. He's a little wary.

When the trawl comes up, it's so full that it takes a winch to get it on deck. The relief is palpable, and so is the excitement. The men gather on deck—scientists, lab technicians, and crew alike—heedless of the rain and lightning, eager to see. This exultation in the big catch is an ancient, primal joy, one that must have touched fishermen throughout time.

It's mainly spot and croaker, two common gulf fish, but there's a total of seven species. Ludsin and his team are weighing them with a hand-held scale, faces scarlet with exertion. Sixty-two kilograms of spot, 113 kilograms of croaker. Ludsin's green shorts are soaked, his hair and glasses drenched.

"That's a lot of freakin' fish," he says, a huge smile breaking out across his face.

We bag some samples of the fish for the freezer, labeling them carefully so Ludsin can examine them in greater detail back in Michigan. By 12:30 he is standing at his computer again, absorbed in plotting the coordinates of the next route. Fish scales—translucent, iridescent—fall from his hands as he types.

*

The fourth day at sea is marked by a sort of tragedy. The cable connecting one of Jochem's instruments to the ship has snapped, and the instrument is lost at sea. Ludsin is on the run, out to the deck. By the time I get there, a ring of five men is huddled around the hand winch, mouths set in a straight line, silent. It feels like a hospital room where the doctors have lost a patient they thought they could save. There is sorrow, grief, maybe resignation. Such are the perils of ocean study.

The sunset sky is lit by the burn of a huge natural gas flare from a petroleum rig off to the side. On the other side, a rainbow.

Roman shakes his head and turns to me. "Everything you put over the edge is expendable," he says.

Now the men converge around Ludsin, the chief scientist, who will have to decide whether anything is to be done. Jochem and Lavrentyev face him, arms crossed, legs set wide, talking earnestly. But it's no use: the *Pelican* can't turn back. Ludsin figures it would take two days to find the instrument at the cost of $6,000 a day to run the boat, plus all the lost research time. The instrument is only worth $15,000. He has to put the expedition's success ahead of the loss of one instrument.

Jochem and Lavrentyev talk with each other for a while. Then they root around and find some unused dumbbell weights on the lower deck of the ship. Somehow they rig up a proxy for the instrument they've lost so that they can keep pumping surface water into the FlowCam, collecting their bits of information.

*

About halfway through the cruise, the order is beginning to fray. Lots of cruises last for five or six days, some for only two or three. Ours, at eleven days, is stretching the limits of human nature. I've read about trips that went on for weeks or months and wonder how the scientists managed.

Like Roman and me, most of the scientists and their team members are working two four-hour shifts a day. When you include the time it takes to get meals in the galley (at the universally fixed shipboard times of 6 a.m., noon, and 6 p.m.), it means everybody's sleep is broken up into three- or four-hour chunks. In our four-person cabin deep in the hold, we're all on different shifts, and we keep waking each other up inadvertently when we stumble into bed in the dark. Ludsin is the anomaly: no one can figure out when he sleeps. Add the general sleep deprivation to the repetitive tasks, absence of e-mail and phones, exceedingly close quarters, lack of privacy, and an absolute prohibition on alcohol, and you have the makings of some short tempers.

For many on this trip the crankiness plays out in relation to food. Food becomes absurdly important on a cruise. Many scientific ships pay as much attention to the care and nurturing of the chef as they do to the chief engineer. The *Pelican*, alas, is having a touch of trouble in this department. Our chef is a last-minute stand-in who's never worked with this group before. He's here on sufferance.

To me, he's a marvel who can throw together fresh and inventive meals for twenty-plus people. But to some of the *Pelican*'s crew, he's a disaster who cooks the wrong things and sometimes commits the maritime sin of putting lunch on the galley table five minutes later than scheduled. The men haven't been getting their quotient of fried foods and it's become a rallying cry of discontent. So now the captain has ordered the chef to pull out the ship's deep-fryer and use it. The galley—which is the only public space on the ship, where we eat, watch television, and read if we can—is filled with greasy smoke.

Roman, the ship's storyteller, has seen it all before. On some cruises the no-alcohol rule flushes out the closet alcoholics who are forced to go cold turkey; then the fried food becomes a salve for the detoxifying liver. Roman remembers one fellow from years back who would change personality and lose twenty pounds as he dried out on each year's cruise.

Roman, who's had his PhD for more than thirty years and is the director of the Horn Point Laboratory, has the long view. And not just on shipboard politics—on the science, too. He's seen the science of the ocean change dramatically in the span of his career. Until fairly recently, he tells me as we monitor ScanFish, scientists understood what was in the water only from what they could catch in nets. And the main reasons they studied the ocean were, first, to catalog what was there and, second, to make sure fisherman could keep catching it. Their motivation was more economic than biological. I remember that Ludsin's former organization, the National Oceanic and Atmospheric Administration, is part of the U.S. Department of Commerce, and it starts to make sense.

Interest in the El Niño weather patterns, Roman explains, developed because those patterns affected the fishery. Early study of plankton came from the same motive: if you could understand when and where the plankton were plentiful, you would know where the fish would be. Then there were all the military reasons for studying the ocean. The U.S. government poured money into ocean research during World War II so that the generals and admirals could understand currents for warfare purposes and retrieve military secrets from sunken submarines. On both sides, that was the main reason governments supported research into submersibles.

So there was little scientific work that pulled together all the pieces of today's ocean system, and even less that compared it with past or future ocean systems. There was no reason to put the present into context. There weren't the tools—taking ice cores, for example—to reconstruct the earth's past environment or predict what could happen in the future. And no one knew that might be important.

But the switch in the ocean system has also caused a switch in the way ocean scientists think. All of a sudden it's become critical to look at the past and the future in order to measure change in the system.

How is the concentration of carbon dioxide in the atmosphere affecting the ocean? I ask Roman.

The surprise is pH, he says. As the ocean absorbs extra carbon dioxide from the atmosphere, its reading moves from the basic side of the scale toward the acidic side through a fairly simple chemical reaction. At the same time, the temperature of the global ocean is on the rise. The ocean's physical structure—the system of currents—appears to be shifting, and with it the pattern of what lives where. All of this affects the food web in the ocean, which is part of what we're studying in microcosm here in the Gulf of Mexico. Scientists don't yet know the trigger points that would push the ocean into a new regime where change would happen exponentially rather than gradually.

And what of plankton, I ask. Will this vast ocean change so dramatically that plankton, the source of half the planet's oxygen, can no longer survive?

Roman chuckles. Plankton, some with calcium shells, some covered in silica, and others naked, are among the planet's most adaptable creatures. They are known to live in waters as cold as 2 degrees Celsius and as hot as 45 degrees Celsius. There will always be plankton in the sea, he says.

I wander across the ship to the lab where Jochem is working. He gained his PhD in 1990 in the great German oceanographic center of Kiel University and worked there for seven years as a research scientist before coming to North America. He's a generation younger than Roman, yet he too has seen vast changes in the science.

He loves the tiniest plankton, called microplankton. Now that their machine is working properly, he and Lavrentyev, who studied in Russia, have been telling me passionate tales about the beauty and diversity of microplankton—their shapes, their colors, and their sneaky strategies to stay alive, which remind me of an intricate war game.

Some plankton are both plant and animal, switching back and forth as needed. Some hunt and chase their food. Some are equipped with needle-like appendages that enable them to puncture other plankton and suck out all the cytoplasm for food. Others build fortresses around themselves using sand particles they capture from the sea; they glue the sand together and then attach spines from other microscopic creatures to create an impenetrable ball.

When Jochem wanted to study these tiny plankton, even some of the giants in his field said they weren't worth the effort. It was a struggle for him to persuade his advisors that he should be allowed to do his graduate work on the subject. Today that attitude is changing, he says. But the science on microplankton is still young. Scientists have only begun to examine the who's who in this planktonic world and are not really looking at which plankton is doing what. Yet these creatures are at the base both of the ocean's food chain and of the chemistry of earth's atmosphere.

One type of microplankton—cyanobacteria or blue-green algae—are the first creatures known to have existed on the planet and have been found in fossils from 3.5 billion years ago. Cyanobacteria are the original photosynthesizers, the source of the planet's first big pulse of oxygen. These tiny plankton thus shaped the very development of life on the planet, which, as our study in the gulf is confirming, could not exist without oxygen. In conducting photosynthesis, in which the energy of the sun is converted to food and oxygen is released, plankton also absorb carbon dioxide from the atmosphere and store it in the ocean. This is one of the planet's main mechanisms for removing carbon from the atmosphere. In other words, the planet's carbon cycle—and, therefore, its climate system—works through plankton.

Just how will these plankton be affected by the oceanic shifts brought on by climate change? It's a big discussion, Jochem tells me, and no one is sure, although it's plain that plankton have adapted to the cycles of the planet and the climate many times through the ages.

One of his concerns is that higher water temperatures might mean less mixing of deep, nutritious waters with the higher levels where the plankton need to live in order to catch the sun's rays. In other words, the plankton may be cut off from all the minerals and food they need to reproduce. (Of course, the opposite could happen, sending the plankton into overdrive, as has happened in the California current.) If that happens, Jochem believes it will be the very tiny-celled phytoplankton that thrive, because smaller plankton have more surface area relative to volume and it's the surface that absorbs food—in other words, they can survive more handily in water that has fewer nutrients. But that has implications for the productivity of the whole ocean. It takes about 100 kilograms of phytoplankton to make about 11 kilograms of copepods, which in turn make one kilogram of fish. When the plankton get smaller, more steps are needed to convert the energy of plankton to the energy of fish. That means fewer fish.

Jochem shrugs. It's all quite speculative, he tells me. No one is really sure what the plankton will do.

*

Stu Ludsin is baffled. He keeps trawling. He keeps getting nothing. One of the other scientists jokes that they're trawling for water. In the dead zone of Lake Erie, where Ludsin has also done a great deal of trawling, he finds that fish collect on the edges of the zone and above it. That's what he's expected here in the gulf. Instead, the fish appear to have fled the area entirely. He calls it "bugging out"—an emergency evacuation.

But his acoustic device is showing a green water column, meaning that something is there, as opposed to a blue column, which would signal the absence of life. "What's the green?" he keeps asking. Is there any life down there? It remains a mystery.

Tonight is the full moon, and that, combined with sleep deprivation, is making everyone a little loony. We're drawn to the deck, paying homage to the heavy, bright moon, posing for pictures in front of it.

Roman begins calling Ludsin "Ahab," after the captain in Herman Melville's novel who single-mindedly pursues the elusive white-backed whale Moby Dick. Somebody mockingly suggests we extend the cruise for another week so that we can trawl some more. Things are a bit out of control. We've been cooped up too long and those narrow bunk beds we will be heading to after this shift are starting to feel like coffins. Last night I shone a flashlight onto the bottom of the bunk just centimeters above mine and found some desperate, inchoate words carved into the plywood—ghost messages from an earlier cruise.

What we don't talk about here on deck is that these scientists love the wonder of it all. Science is supposed to be about solving mysteries, answering questions, laying order on messiness. That's the scientific way. It is, as the great physicist Richard Feynman said, about the pleasure of finding things out.

Yet tonight, giddy under the power of this ancient moon, it seems that we are giving a tip of the hat to the larger order, the unknowable one: the great, plastic, endlessly creative nature that will instinctually find a new balance when the old balance is shot. It's exhilarating for me to stand here, between what's known and what's not, and to see that, try as we might, we may never understand either the old balance or the new. In addition to our ravenous quest for answers, there's a quiet chuckle over the fact that we may never solve all these opaque planetary riddles, that we may only ever succeed in getting a glimpse of what the true questions are.

We abandon the moon, cram ourselves back into the airless ship and seek our berths for a few hours of restless sleep. Alas for Ludsin, the name Ahab sticks.

*

Oxygen at last. We've moved to a different part of the gulf, outside the dead zone, and by 1 p.m. the oxygen has hit 3 milligrams per liter. Ludsin throws his trawl to the bottom and comes up with fish. Lots and lots of fish. Their mouths and gills open wide and settle into the rigid grin of death. They produce copious amounts of slime as they struggle to hold onto life.

"Oxygen, fish," says Roman. "Works every time."

One of the bottom trawls today catches twenty-five different species. Ludsin's graduate students—who seem to need little sleep but tens of thousands of calories of food every day—happily identify the fish with reference books and bag the catch, sprinting to the top deck to throw the sealed plastic bags into Ludsin's chest freezer. Apart from Boicourt's lab technician, who has rebounded from her terrible seasickness, they are the only ones who seem to have energy.

Ludsin tells me that the big test will come tomorrow, when he trawls in a different part of the water column where there's oxygen. That, he says, will show conclusively that the nets are working. So far, the midwater trawls haven't come up with much.

By 7 p.m. we're so close to one of the oil rigs that our mobile phones can hook into their aerial signal. Everyone floods to the upper deck, phones in hand, eager to take up their real lives for a few minutes. My hands tremble as I punch in the numbers that will let me talk to my children and husband.

It's peaceful, hot, wondrous up here on the top deck. As I talk to my son, I see dolphins leaping on either side of the *Pelican*. This doesn't feel like the blob we've been cataloging. Here, outside of the dead zone, there is life.

*

ScanFish is saved by a miracle.

Boicourt, unaccountably anxious over the instrument in the early hours of this morning, happened to notice that three of four screws holding a critical plate were gone. The fourth was hanging by a thread. This is the plate that holds the cable to the instrument. Had the fourth screw failed, the ScanFish would have been lost to the deeps on its next dive. By the time Roman and I show up for our first shift he's screwed the plate back in place and we go out to inspect the repair.

ScanFish is one of about three such instruments in use in the United States and is so old now that it would be hard to replace—that's if the lab could find more than $100,000 to buy a new one.

"We dodged a bullet there," Boicourt says. He and Roman gingerly put ScanFish into the water. It's an act of faith. They rush back to the controls and screens to see how it's doing. Perfect. Better than ever. Even though he's been up since before 4 a.m., Boicourt waits until his assistant, Carol Derry, shows up for her shift at noon. She listens intently as he recounts the narrow escape in blow-by-blow detail. Later they stand together on the deck, watching ScanFish fly under the water, arms gently around each other's waists, wordless, eyes fixed on the cable.

*

The Ahab phenomenon is holding and Ludsin is disconsolate. He's trawling in the middle of the water column where there looks to be oxygen, and still no fish. Just jellies. Again, he worries that the nets aren't working well. He's hoping the night trawls will be better.

Now it's near midnight and the next trawl is due to be put in the water in a few minutes. Ludsin is standing at his computer in his rubber boots, a huge stack of papers to one side. He's making notes on the draft of a student's master's thesis. He's pale and has dark shadows under his eyes.

When the trawl comes up it contains lots of triggerfish, a popular aquarium species. It's not what he was looking for, but it tells him that the net has the capacity to catch fish . . . if only the fish were there.

*

Finally we're heading back to dock at Cocodrie on a continuous tracking route that will take us two days and cut through part of the blob from one end to the other. The other routes have been small slices, tracking in and out of the dead zone and along its margins.

By this time Roman is looking like a younger version of Clint Eastwood, stubble covering his chin. Boicourt is immaculate, patrician as always. Ludsin just looks exhausted. He hasn't spent much time on his bike.

We're talking about a heretical idea, another key scientist's tactic that can sometimes help frame the right questions. Sure, there's a blob. It's big at 17,000 square kilometers and has doubled in area since it was first routinely measured in the mid-1980s. It's unmixable except by hurricanes. It seems to be devoid of large forms of life; the general panoply of life you'd expect in these waters—including bottom-living one-celled creatures—is also absent. We will find out later, from the trip's results, that the dead zone is completely devoid of fish and of the zooplankton they feed on. The zooplankton that remain are squeezed up into the warmer, light-filled waters near the surface, which still have oxygen. But there they can't hide from the fish, and their eggs sink into the zone and suffocate before they hatch. Even the one-celled protozoa—protists that are neither plants nor animals nor fungi—are reduced in the oxygenless water.

Does it matter, though, in the grand scheme of things? Maybe the dead zone is just a biological black hole that comes and goes. Don't the fish just go to another part of the gulf and carry on their merry way? While fishermen have been forced further out into the Gulf, the value and tonnage of their catch hasn't changed much since the 1970s, when this blob first began to establish itself.

Roman has done a lot of work in the dead zone of Chesapeake Bay, off the northeast coast of the United States. It's clear there that the blob has had a hand in wiping out the sturgeon and diminishing the oysters by taking away the living space they need to spawn and live. And it's definitely had long-term adverse effects on the environment. Studies in other parts of the world, including in the Baltic, Black, and Kattegat seas, show that dead zones there have destroyed fisheries entirely or caused them to crash.

As for the Gulf of Mexico? Is it a signal of much worse things to come, particularly as the climate keeps changing and the fertilizers, now, ironically, being used in even greater quantities to grow corn for biodiesel fuels in a bid to slow down greenhouse gas emissions, keep flowing down the Mississippi? What will the future bring? And how does it connect, if at all, with the great oxygen starvation that happened in the ocean at the time of the Great Dying?

Ludsin reminds me that ours is the first study to look at the whole column of water in this dead zone rather than just the underside of the blob on the floor of the gulf. All we know for sure is that historical fishing grounds are barren during much of the year, that the essential nature of a living resource has changed.

Nancy Rabalais is the marine scientist at the Louisiana Universities Marine Consortium in Cocodrie who discovered the gulf's dead zone in the mid-1980s and has tracked it every year since. She's known as the "Queen of the Dead Zone" in the popular press and has been pilloried by farmers throughout the United States for fingering their fertilizers as the cause of the blob. For years she was a voice in the wilderness in the scientific world for her work on this phenomenon. "It was sort of like acid rain in the beginning. Nobody believed it," she told me in Cocodrie before we set sail. Now, her work has spawned all sorts of other studies, including this cruise and a multilevel government task force recommending voluntary measures upstream intended to shrink the blob to no more than 5,000 square kilometers.

It's a far-off dream.

Even if the plankton-feeding nitrogen and phosphorus were cut down, no one is certain that life in the dead part of the Gulf of Mexico would return. There are dozens of scientific papers Rabalais and others have written on this dead zone; I've been reading them between ScanFish watches.

They report that when the former Soviet bloc collapsed and subsidies for fertilizers dropped during the 1990s, the nitrogen flowing into the Black Sea also fell. By 1999 the Black Sea's dead zone had receded dramatically—but the plankton and fish didn't come back. Nor did they return in the Bodden inlets off the Baltic Sea.

It turns out that the soil acts as a sponge for synthetic nitrogen, and possibly has for decades. A UK study on experimental plots of land showed that they continued to leak significant amounts of nitrogen forty years after it was last applied to the soil. In Sweden, nitrogen kept leaching into the water for thirteen years after fertilization had stopped, and in amounts that did not diminish. So even if the nitrogen eventually washes out and the oxygen comes back, the original mix of life in the former dead zones may not return. As Rabalais says in one of her papers, they may have developed a new, different normal, one that only lightly resembles what was there before.

Rabalais has come to be concerned that humans could be shifting the global nitrogen cycle through the use of synthetic nitrogen compounds in agriculture. If that is so, what effect will it have on the global ocean systems? The answers are as yet unclear.

*

By midnight the constraints of this long trip are loosening. A bunch of the men are gathered in the galley for a monster game of poker, betting with mini chocolate bars. It's the first time this trip I've seen Ludsin sit down. His feet are up and he's eating ice cream out of the tub.

*

We're almost back to shore and it's my last shift on ScanFish. Roman lets me track it on my own for a few minutes while he runs off to do something else. When he comes back, he gives me a high-five. "You flew solo!" he says, grinning.

Then, Cocodrie and firm land appear. I jump off the *Pelican*, legs wobbly, breathing in the earth and trees. The scientific team has been packing up for hours and now all those kilograms of equipment and samples, containing millions of pieces of new data, need to be shifted off the boat and into vans, trailers, and boxes to be posted home.

Ludsin loads the chest freezer into his U-Haul and gets behind the wheel of the van. The freezer holds hundreds of kilograms of fish, hard caught, and he's got an eighteen-hour drive to Michigan ahead of him before he can plug it in again. It will take the scientists years to figure out the meaning of all they've just gathered.

As for me, I have a longing to understand part of the ocean's chemistry that this crowd hasn't been looking at: pH. So I track down two of the world's experts and fly to Puerto Rico to meet them at a scientific conference.

Reading the vital signs: pH 3

PUERTO RICO When the seas turn to acid

Every single reaction that makes up the chemistry of life occurs in water. Water, the universal solvent, dissolves almost anything, including gold.

Chemically, water—the most abundant molecule on the planet—is made up of two hydrogen atoms and a single atom of oxygen. And one of the defining chemical characteristics of water is its pH. Pure water at room temperature has a pH of 7. Above that level, the water is basic; below, it is acidic.

The creatures that live in the ocean are extremely sensitive to pH and do not thrive when ocean chemistry changes. Fortunately, the pH of ocean water is not easy to change. For millions of years the pH of the salty water of the open ocean has remained stable at 8.2, a level tied to the concentration of carbon dioxide in the atmosphere. When a change does occur, there are elaborate systems in place to return things to normal, mainly through the weathering of rocks from land. It takes something cataclysmic to effect a lasting change.

When masses of carbon dioxide gas are introduced into the atmosphere through slow planetary processes spanning a million years or more, such as the shifting of tectonic plates, the ocean holds its own. But when carbon dioxide gas is thrust into the atmosphere quickly—meaning, in this context, in 10,000 years or less—ocean pH falls. The water becomes more acidic.

Ken Caldeira and Michael Wickett, who model deep-time earth systems at California's Lawrence Livermore National Laboratory, are extremely good at explaining this, and did so in a 2003 paper in *Nature* magazine. But they were curious about the pace of the increase in carbon dioxide concentration in today's atmosphere compared to what had happened over the past 300 million years. So they ran some models. They calculated that if humans keep burning fossil fuels at the rate we have been in recent decades, until we burn up the last of them in centuries to come—an extremely rapid infusion of carbon dioxide, in geological terms—the ocean's pH will fall by 0.7 of a unit. Geological records show that the ocean's pH has never in 300 million years dropped by more than 0.6 of a unit, with the possible exception of some "rare and catastrophic events in earth's history." The ocean's pH, it seems, is not something to trifle with.

Joanie Kleypas and Chris Langdon know this to their cost. We are at the end of a groundbreaking scientific meeting in Puerto Rico and we have finally found a few hours to talk. Kleypas, a scientist at the National Center for Atmospheric Research in Boulder, Colorado, and Langdon, who is at the University of Miami, are slightly wary as we sit down together, in the way that scientists who are delivering profoundly unpopular findings can sometimes be. It's not that they don't trust their findings; it's that they have been so thoroughly spanked by other scientists whenever they try to explain them. Their conclusions are outside the canon.

This morning was a case in point. Kleypas and Langdon made a pithy presentation to their fellow scientists, arguing that a new monitor being installed on the coral reef in the bay in front of our hotel ought to capture information on the pH of the water and how the corals are responding to it. It was a hard sell, followed by harsh questions, even from this intimate group. A much harder sell, for example, than the pitch from a fish biologist to install an acoustic phone so that scientists could listen to the fish chatter and hope for clues about how healthy the coral reef is.

The heresy of Kleypas and Langdon is stating that carbon dioxide entering the atmosphere from burned fossil fuels is affecting the global ocean's acidity and, therefore, the proportion of carbonate ions in the water. The proportion of carbonate ions in turn affects the ability of creatures to use the calcium in seawater to construct shells and skeletons. Further, they say that by the time the carbon dioxide concentration in the atmosphere reaches double its preindustrial level (560 parts per million, versus 280 in 1850)—which could easily happen by the middle of this century—marine organisms will have access to 30 percent less calcium that they once had. By the time it triples, there will be just half the available calcium.

It's a doomsday scenario for corals, some shelled plankton, and many other creatures that depend on calcium to live—they're known as calcifiers—because at some point they won't be able to build their shells and skeletons. Eventually, when the seas become acidic enough and the concentration of carbonate ions drops low enough, these creatures' calcium shells and skeletons will simply dissolve.

Kleypas remembers when she first recognized the seriousness of the pH issue. She was in a meeting with Langdon in Boston in 1998. A committee of academics affiliated with the Society for Integrative and Comparative Biology, once known as the American Zoological Society, was talking about carbon dioxide and how it affects the concentration of carbonate ions in the ocean, a topic so arcane and seemingly irrelevant that, Kleypas remembers, the group's members hadn't even heard of it at university. It certainly wasn't on the scientific radar. Nevertheless, they pooled their information and did some back-of-the-envelope calculations. Once they figured out how low the carbonate ion concentration would fall if carbon dioxide concentrations in the atmosphere kept rising they realized they were looking at a marine Armageddon. Kleypas ran into the bathroom outside the committee room and threw up.

"It's pretty serious when you're messing around with calcifiers," she says, explaining that the fossil record shows that when carbonate saturation falls, some creatures that need calcium become extinct. Would humans survive if coral reefs and calcium-shelled plankton and other sea creatures that need shells disappeared? How much of a life support system are they for us?

Plankton produce half the oxygen in the atmosphere and are at the base of the planet's largest food chain. When they make calcium-based shells and when they make energy from the sun, they're storing carbon in the ocean, keeping it out of the atmospheric system. (Not all plankton make calcium shells. Some have no shells and others have shells of silica.)

We have no idea what other life forms that we depend on would vanish if plankton were to die out or dramatically decline in number. And we still don't know to what degree humans are symbionts with all these creatures we are endangering. Kleypas and the others who are working on this issue of ocean chemistry have unwittingly embarked on the ultimate human quest: to find out not where we came from but where we are going. It is the great endpoint of human imagination.

*

A year later those offhand calculations from the Boston meeting became a major article in *Science* magazine, one of the bibles of the scientific world. They received a lot of press coverage and drew a chorus of outrage from fellow scientists who questioned everything about the premise, from the chemistry to the calculations to the assumptions.

The ocean, the critics contended, is too robust, too massive, to be affected by human actions. It has backup systems for its backup systems.

They called for more investigation and more analysis, including examination of other eras in the earth's history when ocean chemistry has changed rapidly. All subsequent research has backed up Kleypas, Langdon, and their colleagues.

What does all this really mean? Why is it so important that a top scientist would vomit when she figured it out? It's because she understood the ineluctable poetry of chemistry. In plain language, it means that when we pump carbon dioxide from the oil, gas, and coal that we burn into the atmosphere, it also enters the ocean. The two media of life on the planet—air and water—exchange gases across the vast surface of the ocean, exactly where the water molecules meet those of the air. It's a continuous, two-way process that is key to the earth's ability to support life as we know it. When we add carbon dioxide from the oil, gas, and coal that we burn to the atmosphere, this gas exchange alters the fundamental chemistry of the ocean. That's in addition to changing the climate—a separate and also severe problem.

Of all the carbon dioxide that cutting down trees and burning fossilized carbon has added to the atmosphere, roughly a third has remained in the air, a third has ended up in the ocean, and a third has been reabsorbed into the land. In the atmosphere, carbon dioxide acts like a giant blanket, trapping heat over the planet. It remains chemically boring, not reacting with anything. It just hangs there, intact, making up an ever greater proportion of the air. Today, that proportion is approximately 387 parts per million by volume, compared to about 280 before industrialization began. There's no scientific evidence that that number has risen above 300 parts per million at any other time in the past 20 million years, but now it's increasing exponentially each year and is expected to reach 560 parts per million, or double preindustrial levels, by about 2065.

In the ocean, there's already so much human-generated carbon dioxide that scientists have tracked it, with its unique chemical markers, to the depth of 1,000 meters. In the cold and restless North Atlantic, where waters churn in a great, deep conveyor belt, scientists can measure the human carbon fingerprint at 3,000 meters.

This is an almost unimaginable amount of carbon.

Carbon dioxide in the ocean is not inert as it is in the atmosphere. It metamorphoses. Some of it dissolves and interacts with other substances, including sea water. This produces a lot of electrically charged atoms, known as ions. These ions have either gained electrons, meaning they are negatively charged, or lost them, meaning they are positive. Like magnets, these charged atoms either repel or attract other ions, leading to chemical reactions. Carbon dioxide added to seawater produces positive hydrogen ions plus several negatively charged forms of dissolved carbon, including the carbonate ion—the one that Kleypas is so concerned about—and bicarbonate.

The more carbon dioxide the ocean absorbs, the more chemical reaction occurs. And the more chemical reaction, the more positive hydrogen ions. The more hydrogen ions, the lower the pH. The more carbon dioxide the ocean absorbs, the less basic and more acidic the ocean becomes.

So far, pH has dropped by a little more than a tenth of a pH unit, from about 8.2 to about 8.05 in the surface of the global ocean, meaning that it is still slightly alkaline. This drop may not seem to be significant, but in terms of how it affects biology, it is huge. That's because pH is measured not on a linear scale, like a yardstick, but on a logarithmic one. Going up the scale, every unit is ten times as large as the unit that came before. Going down, each unit is only a tenth of the unit above. Think of a cyclone, small at the bottom and cavernously wide at the top.

The plants, animals, and bacteria that live in the ocean have evolved to thrive in the pH that existed before humans started fiddling with it. They're used to chemistry that stays within a certain range.

Langdon, sitting here in the tropical heat of southern Puerto Rico, tells me about two cruises that ran from Antarctica to Iceland, stopping every 30 miles to check pH. One took place in 1991 and the other in 2005. Even in that time, pH had dropped.

Again, it's helpful to think of the human body. The pH of arterial blood in humans has to stay between 7.35 and 7.45. If it goes higher or lower than that we get very sick and could die. Already, ocean pH has dropped by an amount larger than the range of healthy pH in the human body. And it's still changing, as the carbon dioxide load in the atmosphere keeps increasing. Despite political attempts around the world to lower carbon emissions, they are actually rising more quickly now than they have at any time since we started measuring. A paper published in late 2007 found that global emissions from fossil fuels and cement making, which entails burning limestone, were more than a third higher in 2006 than they had been in 1990.

*

Earlier this week, before we got mired in heresy and pH, Kleypas and I did some snorkeling on the coral reef here at La Parguera Natural Reserve.

Because Puerto Rico is a territory of the United States, this reserve is part of the American network of protected marine areas. Its main attraction is a bay that lights up the dark with the natural fluorescence of marine life. It's become an international tourist draw. More than that, though, this reserve has a lot of the ocean habitats that characterize the tropics, and all easy for researchers to get at.

Among them are the mangroves, with their complexes of roots that can turn sea into land. Because life collects at the margins where different habitats collide, mangroves are extraordinarily rich, feeding and sheltering creatures that depend on air and water as well as land. Some of the mangroves are making tiny islands, the promise of new earth to come. There are wide beds of sea grasses, plants whose ancestors made the leap from ocean to land and then went back to the ocean again, as well as vibrant shorelines, ponds crusted with salt, and flat plains thick with the larger algae.

The reserve is also home to a coral reef, which forms protective rings along the outside of the calm bay overlooked by our hotel. Just outside the reef, the island's placid shelf drops abruptly to several thousand meters, opening up to the mysteries of the deeps. At the heart of the whole reserve is the University of Puerto Rico's Magueyes Island Laboratory, a short boat ride from our hotel.

We had taken a boat out to a section of the reef where technicians and scientists were struggling to put up a monitoring tower that would provide a continuous stream of information about the sea. It was the third such site that the U.S. government had set up, and more were planned in other parts of the ocean in the months ahead. The session we were taking part in was aimed at advising the government about what information ought to be collected. I wondered why there hadn't been more data collection before this. Hadn't there been enough evidence of change to warrant at least looking at the numbers?

So much of the sea is invisible, on some level a notional place rather than a real one, that it's been easy for humans to ignore. And it's now up to those, like Kleypas and Langdon, who have figured out what's really happening to make its watery darkness visible to the rest of us. Snorkeling in this protected place, I thought of the arc of human knowledge about the ocean over time; it's like the shift in pH, slow at the beginning but expanding exponentially in recent years.

As far as scientists have been able to piece together, humans first settled by the sea only about 164,000 years ago, a blink of the eye given the

4.5-billion-year history of the planet. The evidence comes from a recent archaeological find at Pinnacle Point on the South African cape: a cave filled with the remains of cooked shellfish. Until those ancient African ancestors settled on a cape beach, our species, which had evolved less than 40,000 years earlier—a latecomer to the planet—knew the plains and the jungles but not the vast ocean or its shores. Scientists concluded that those early shore-dwelling humans made symbolic use of the mineral pigment ocher, an important sign that they were capable of abstract thought, a leap forward in human evolution.

In terms of nutrition and transport—not to mention climate and oxygen—human civilization and the sea are intimately connected. Many evolutionary biologists have theorized that humans were able to develop intellectually and travel so far from our African birthplace because we discovered seafood, with its high fat content—the perfect brain food. That model draws a direct line between the sea's riches and what humans are today.

Those early humans who glimpsed the ocean off the cape could have had no concept that water covers nearly three-quarters of the surface of the planet. They could not have known that the ocean was the cradle of life or that humans develop in the salty water of the womb with gill arches and gill slits that later develop into parts of the ear. They could not have recognized the ocean as an integral part of the carbon and oxygen cycles, or even have known that such cycles existed. Still less could they have imagined that life as they knew it was dependent on the sea staying healthy and stable.

Even much later, when the Mesopotamians started describing our planet for posterity, they imagined a flat disk bathed by a salty body of water. To these thinkers, the planet was all about land. Indeed, the English name for our planet, earth, comes from the Anglo-Saxon word for soil or land.

The idea that the planet was mainly land has persisted. The first, highly metaphoric maps of the world show land taking up most of the surface of the globe. The marvelous maps of Claudius Ptolemy, who lived in the second century, show a preponderance of land and little water. In his day, as throughout much of human history, the intellectual quest was to figure out how that land-planet interacted with the heavens, not how land, air, and sea were conjoined. It was thought that the stars controlled the destinies of individuals, societies, even empires, with the heavens—or their rulers, the gods—occasionally intervening in human affairs to punish the wicked and reward the good.

In the early Middle Ages, mapmakers put Jerusalem at the physical center of the planet. Land masses still covered the bulk of its surface, with tiny rivulets marking the rivers and seas as if they were watery roads. And while there were inklings by the sixteenth century that the planet was mainly water, the sea was still largely perceived as a two-dimensional surface. It might provide fish and other food, but its primary use was for sailing to other places, for projecting military might and trade dominance. The seas were named after the lands they abutted, and the names given to many sea creatures—sea horse, swordfish, sea cow, catfish—liken them to terrestrial creatures and goods. The immense third dimension of the ocean—the vast deeps that give it its biological importance on the planet—was still unimagined.

When humans finally began to explore those depths systematically in the eighteenth and nineteenth centuries, it was mainly to make shipping safer. In 1876, when the British scientists aboard the *Challenger* sailed home with their findings, people learned that parts of the ocean floor were far deeper than anyone had imagined. But it was not until the 1930s that anyone physically went to the ocean's depths. In the 1950s such expeditions became more commonplace, motivated largely by the military goals of recovering sunken ships and missiles and laying cables. The deepest point in the ocean, some 36,000 feet in the Pacific Ocean's Mariana Trench—greater than Mount Everest's height above sea level—was identified in 1951 and first visited in 1960. It is ironic that humans have been drawn to the depths by the promise of lost treasure or other artifacts of human life on land, while mainly remaining unaware that the ocean itself is the real treasure.

Until quite recently biologists have thought of most of the ocean as a desert—and an immutable one at that—unaware that it teems with life even in its deeps. It's only in the past several generations that they have come to understand the importance of the ocean to life on the planet as a whole and realized that the ocean's life forms have changed dramatically since the beginning of time, in step with pulses of extinction and new creation. (Jacques Cousteau was one of the early underwater scientific explorers, finding creatures that had never been seen before by diving with an air suit.) The idea that humans can change the ocean's very chemistry and its capacity to support life is newer still. That the ocean contains the switch of life, and that the human hand is on that switch, is a concept on the frontier of scientific thought.

*

At La Parguera, where technicians were installing the monitoring tower as Kleypas and I snorkeled, the reef was in poor shape. Just 20 percent of the coral that ought to have been there was still alive. But it still had beauty and majesty. It was a good reef to experiment on, to see whether it was possible to put the pieces back together once the whole had been destroyed.

Some researchers were trying to rear baby corals. Others were attempting to seed a type of sea urchin, *Diadema*, that used to eat a lot of algae that no other creature ate. Sort of like the bison on the North American plains, the urchins were the main grazers of the coral reef, keeping the algae in check so other animals and plants had room to grow. In 1983, before most scientists had discovered the critical role the sea urchins played in maintaining the balance of the reefs, the urchins disappeared mysteriously right across the Caribbean. And while they've started coming back at a glacial pace in the past few years, they are still too rare to keep the algae down.

I put my arms at my sides, kicking against the waves, moving with the water, making progress toward the shore, eager to see what was there. Closer to shore, the reef was a graveyard. Only about 5 percent was still alive. I kicked up against a type of *Acropora*, one of the branching corals that helped build this reef, and recoiled in case I was hurting it. When I went back to look, I saw it was dead. Just rubble.

There's a long list of hazards for corals in the Caribbean, even before the big effects of changing ocean chemistry kick in. Over the past three decades, roughly 80 percent of corals in the Caribbean have perished. Some have died because the carbon dioxide we're pumping into the atmosphere is hiking the temperature of the water, killing the symbiotic algae that live inside the coral animals and feed them. Some are dead because overfishing across the Caribbean has removed some of the creatures that kept marine life balanced. Others have succumbed to the sewage and pesticides that wash down from humans on the land, or to dust-borne fungal spores, disease, and toxic chemicals that have blown in from Africa in increasing amounts since drought has set in there.

Some have fallen to a strange and virulent outbreak of coral diseases that have infected the Caribbean. With only 8 percent of the world's coral reefs, the Caribbean is reporting 66 percent of the world's coral disease events. Some corals are being sickened by four or five diseases at the same time. A leading theory is that the cumulative effects of all the environmental stresses have left them unable to defend themselves against illnesses.

All in all, the corals here are dying faster than they can grow back, even in this protected area. The situation, as one of the coral scientists has said, is not good. It's almost as if the switch to a new system has already happened here. A new balance may be on its way, one that will reinforce itself rather than trying to re-create the old system.

Kleypas and I kept swimming. Finally we found an *Acropora* that was still alive. Just one, in a field of crumbling skeletons. It spread like the rack of an elk, brown, regal.

*

The interview at our parador is drawing to a close. The late afternoon sun is hot and clear, the ocean heartbreakingly blue, invisibly ill. Some of the other scientists from this conference are cavorting at the pool, kicking off the intellectual rigors of the past several days.

Kleypas and Langdon could probably talk about this for many more hours, but I'm spent. The implications of what I've been hearing snake hundreds of thousands of years into a future that we can only try to imagine. I'm overwhelmed. We still have only pieces of this puzzle, and the picture they are showing us is terrifying.

How do these scientists on the front lines keep going? Langdon points to the great strides humans have made to stop acid rain and heal the Great Lakes, the banning of DDT sparked by Rachel Carson's book *Silent Spring*, the attempts to repair the ozone hole.

Kleypas has hope, too. But, like Langdon, she is an intelligent optimist, not a faith seeker. She has a place in Costa Rica, the Central American country fabled for successfully basing its economy on conservation rather than destruction. She stays there for part of each year.

The rainy season at the end of 2005 was unusually long and wet in Costa Rica for reasons unknown but possibly connected to global climate change. The jungle trees there don't flower until the rain ends. But the rain was so intense that the many of the trees didn't bear fruit—the rain knocked off the flowers. That small change meant that as many as half of the spider monkeys in the country's Corcovado National Park starved to death. At the end, the monkeys were so ravenous that they came down from the trees to scavenge for food—unsuccessfully—on the jungle floor. Nobody had expected it. At first, biologists thought the monkeys were diseased, maybe with yellow fever. Finally they realized that the timing of the food cycle was off. That same year, baby birds on the Great Barrier Reef had died because the fish there moved to cooler waters, and at Mon Repos on the Australian coast unborn turtles cooked in their own shells.

Kleypas's point is that we only feebly understand how some of these great earth systems work. The planet remains capable of tossing us some big surprises. In the ocean, we understand even less. We don't comprehend enough to know how great the risks are. In fact, the further along the logarithmic index of knowledge we get, the more we understand how little we know.

There are only a few places in the world where scientists are putting the theory of ocean acidification into practice. One is Plymouth in England; I've run into some of the scientists working there in the course of my research. It seems critical to me as I leave Puerto Rico to go on to England and see what they're finding out. When I call them up, they're only too eager to show me around.

*

I check back with Kleypas in 2008—two years after our first meeting—to ask whether the monitoring system at La Parguera is tracking seawater chemistry as she and Langdon had suggested. By that point they have done some testing of water off the reef and found that despite the poor state of the reef, they can still get a little information about how the reef is conducting photosynthesis, breathing, and forming calcium. There's only just enough of it left.

Reading the vital signs: Metabolism 4

PLYMOUTH, ENGLAND

The fate of the plankton

Plankton are the nonchalant wanderers of the ocean. They go with the flow. And one of the toughest, least understood questions about the changing global ocean is where plankton will "go" next—in the evolutionary, biological, chemical, and even metaphysical sense, as well as the geographical.

Does it matter where they go?

Resoundingly, yes. This question, though unimagined by most of us, may be the most important question humans will ever grapple with.

Plankton, a group of tiny organisms that range from microscopic marine viruses and bacteria to single-celled plants with fabulously ornate shells to minuscule plant-eating animals, are the lynchpin on which life itself depends. Not only do they produce half of the planet's oxygen but they are also responsible for a host of different, invisible, and interlocked parts of the metabolism of the planet. These aimless drifters reside at the very bottom of the ocean's food chain and at the very top of the chemical cycles—oxygen, nitrogen, and carbon—that support life on earth. It is fair to say that if plankton vanished tomorrow, the marine food chain would fall apart.

Despite their immense planetary significance, hardly anybody gives plankton a thought, even in the scientific community. One notable exception is the clutch of scientists who do rough-and-tumble plankton experiments at the marine institutions of Plymouth in southern England. The standing joke is that Plymouth has more marine biologists per square meter than anywhere else in the European Union. And without question, Plymouth Sound is one of the best studied bits of ocean anywhere.

To begin with, Plymouth is home to the second oldest marine science organization in the world: the Marine Biological Association of the United Kingdom, established in 1888. Seven of the association's scientists have captured Nobel prizes. The Sir Alister Hardy Foundation for Ocean Science is also based here. Its scientists have been tracking plankton's wayfaring habits for more than seventy years using hand-tooled machines that rely on ancient watchmakers' techniques. It is the longest-running marine biology survey in the world.

As well, the seaside city boasts Plymouth Marine Laboratory, an internationally recognized center for research on marine topics. Rather than looking only at small pieces of the ocean's problems, the laboratory takes an unusually comprehensive view. Its aim is to examine plankton and other marine creatures, using techniques ranging from satellite imagery and computer models to hands-on experimentation, in order to assess the health of the planet itself—and the implications of any changes in its health. Its scientists are on the very cutting edge of understanding the ocean's status as the main life-support system of the planet.

When I encounter several scientists from Plymouth Marine Laboratory at a meeting in China, I notice that they give by far the most bracing and forward-looking talks of the week-long event. If anyone can tell the full story of plankton, this crowd can.

So here I am in Plymouth. Because I'm dealing with plankton, which are so inscrutable, my path of research is not straightforward. It meanders from scientist to scientist, experiment to experiment. The only common thread is that all of the scientists and experiments are in one city. I've come to think of Plymouth as Plankton Central.

The marine ecologist Melanie Austen leads a group of scientists doing experiments on ocean pH. Here, the changing acidity of the global ocean is not controversial, as it was at the meeting in Puerto Rico—it's incontrovertible. Many of these scientists have been responsible for teaching government leaders in the UK about what they call the "other carbon dioxide problem" and helping to put it on the political map.

While Austen and her colleagues have no definitive findings on how plankton will respond to more acidic seas, they are getting clues from the creatures that live on and in the mud of Plymouth Sound. They start by re-creating the ocean in covered vats in their cavernous, frigid laboratory, then add cores of mud containing seabed animals. Finally, they pipe carbon dioxide into the air above the vats and watch what happens to the acidity of the water and to the animals living in it.

As Joanie Kleypas, Chris Langdon, and other scientists have calculated, the more carbon dioxide is pumped into the vats, the lower the pH goes. Many of the animals react catastrophically, while a few, including some types of crabs and mussels, seem to be able to compensate for the lower pH, at least as adults. Those that don't compensate can't maintain their internal chemistry. The intimate processes of the cells—building proteins, using enzymes—stop working as they should. In human medical terms it's akin to the diabetic whose blood sugar rises far too high. The chemistry of the blood is wrong, and that affects the heart, the nervous system, the circulation, and other bodily functions.

A member of Austen's team shows me photographs of the guts of sea urchins and brittle stars (relatives of starfish) that had failed to survive in the more acidic water. Their digestive systems had fallen into shreds. At the same time, their reproductive systems went slightly mad, producing genetic material with frenzied abandon in an attempt to create offspring before they died. It's a well-cataloged animal response to full-body stress. But, as Austen points out, baby urchins and brittle stars—more vulnerable than their parents to the increasing acidity—would die in the hostile new environment anyway. Plus, as the pH continues to fall, the animals' developing eggs start to break down, meaning that they will eventually stop reproducing entirely. This is the picture of an utter failure to thrive.

These animals are engineers, working away on the seabed and helping to control how quickly the food and minerals in the mud there get recycled into the chain of life. A change in acidity affects not only the animal but also its function within the wider system. The same factors may affect plankton, not to mention countless other plants, animals, bacteria, and viruses in the sea.

This situation is what Jerry Blackford, one of the people at the laboratory who runs computer simulations of ocean acidification, calls "the disconnect." His models show that by 2050 the pH will be lower than it has been for twenty million years. The ocean's life forms will be disconnected from their own evolutionary heritage.

Blackford has been studying a plentiful, ancient plankton called coccolithophores. Known as coccoliths for short, these are one-celled marine plants that make complex plated armor for themselves out of calcium, carbon, and oxygen that they absorb from the ocean. Each species of coccolith—there have been about 4,000 since they emerged in the Jurassic period, and most are now extinct—makes a unique type of shell. Under the microscope some of the plates look like the lids of garbage cans, others like the leaves of an artichoke. Many resemble the petals of flowers, sculpted from limestone. The white cliffs of Dover are made up of trillions upon trillions of these microscopic plates of calcium carbonate.

Like corals and a raft of other marine creatures, the coccoliths need calcium to make their armor. The more acidic the sea becomes, the less calcium is available for plates, reefs, shells, and other calcium-needy structures. How would a naked coccolith fare? That's uncertain, says Blackford. One possibility is that they would die off, having become more vulnerable to viruses and predators. But this could become a moot point: once atmospheric carbon dioxide gets to 1,000 parts per million—it is now at about 387—the models show that the number of coccoliths in the ocean crashes, whether they are naked or plated.

These microscopic creatures, so tiny that a level teaspoon could hold about 860,000 of them, are the prime mechanism for storing carbon in the deep ocean. In fact, when they evolved hundreds of millions of years ago, they revolutionized the regulation of ocean carbon chemistry. If the coccoliths can no longer make their plates because there's not enough calcium available, or if their numbers decline due to acidic seas, there will also be a drop in the amount of carbon that they absorb from the ocean and, when they die, send to the seafloor for thousands of years. How will the effects of such a change cascade through the fiendishly intricate ocean system?

Says Blackford, "The first honest answer is that we don't know."

*

I leave Blackford's office and wander along the shore of this port city. My cheeks turn scarlet in the sharp winter wind. The air carries the ancient scent of decaying fish and sea salt.

Plymouth Sound is one of the largest natural harbors in Europe, and that has made it the stage for many important scenes in Western history. Even today, the Royal Navy plays war games in the harbor once a week, on Thursdays, summoning warships from distant points to practice maneuvers.

From this harbor, Francis Drake launched his storied circling of the planet in the sixteenth century. In the seventeenth century, the pilgrims marched down the Plymouth steps and onto the *Mayflower* to make their home in the New World. In the eighteenth, James Cook set off on three expeditions, in a race to discover, map, and claim the world. And in the nineteenth, Charles Darwin caught a fair wind from here on board the *Beagle* for the five-year journey that would change how humans understood their origins and their place on earth.

Humans used to regard the sea as a worthy foe to be conquered by seacraft so that it could be used to get us to other lands, or to concentrations of fish or other humans in ships. Triumph over the sea brought with it the ability to conquer new territory, build empires, get rich, vanquish hunger, and dominate others. Are we capable, now that we have a more sophisticated understanding of the sea, to revise this primal understanding of its meaning? To understand that we need now to protect it instead of forcing it into submission?

John Zachary Young, one of the most eminent biologists and neurophysiologists of the twentieth century, published the definitive textbook *Life of Vertebrates* in 1950. It is still a university staple. Young studied the rhythms of evolution, the great, rapid planetary spasms that created a lot of new species in a relatively short period of time. And not just the individual events or even the larger patterns, but also the underlying causes that determine the patterns. He investigated the workings of the systems that create and destroy new species and sought to understand how one might predict the future workings of those systems. "Too often in the past we have been content to accumulate unrelated facts," Young writes:

> It is splendid to be aware of many details, but only by the synthesis of these can we obtain either adequate means for handling so many data or knowledge of the natures we are studying.
>
> In order to know life—what it is, what it has been, and what it will be—we must look beyond the details of individual lives and try to find rules governing all.

This reasoning took Young to the ocean:

> The composition of the ocean is of special interest to biologists since life first became possible because of conditions in the sea, and then evolved there for perhaps [two billion years], that is to say for the greater part of its whole history.
>
> Living things still retain in their ionic make-up certain characteristics of the sea, indeed some authors have interpreted the blood plasma of vertebrates as a relic of the Palaeozoic sea.

He's saying that the sea is the mother of life and that we carry her within us. We are connected in the most fundamental possible way with the global ocean that gave us life.

This is poetic, but it is also biologically significant. It means that the chemistry of the ocean over time has helped to determine what forms of life could come into being. The ocean holds not just the origin but also the fate of life. So to me it follows that if the chemistry of the ocean is changing now, the types of life that it can support will change too.

Whether life collapses catastrophically, as it has only five other times in its history on this planet, depends on how quickly and how deeply the ocean changes. For example, during the Great Dying of the Permian period, 250 million years ago, when the sea lost its oxygen, about 95 percent of life in the ocean vanished from the global gene bank. Including a lot of plankton.

In the history of human thought, the idea that life could change so dramatically is new. Young, writing long before the current ecological crisis was recognized, noted that it was only in the nineteenth century that scientists realized that the geography and climate of the world had not always been the same, adding:

> But only in the last twenty years has the full extent of the change become appreciated. . . . To understand why animals and plants have changed, and how they have become spread over the earth, we need knowledge of the distribution and composition of the land, sea and atmosphere at various times in the past, as well as information about the temperature and other physical conditions.

In short, chemistry and physics determine biology. And future chemistry and physics will determine future biology.

The revelation now, sixty years on, is that it is our hand on the switch.

That brings me back to coccoliths, and to the scientists' trick of deconstructing the question. In scientific circles, this is known as performing "the big so what."

So the coccoliths, here on the planet for at least 150 million years, won't like an acidic sea, may not produce their limestone shells, and won't store carbon in the deep ocean. *So what?* Does any of that really matter? Won't another type of plankton—or many other types—take over the coccoliths' planetary functions in good time if they vanish? Is this scientific race to find out about plankton really necessary? Couldn't humans just noodle along, having faith that the planetary systems will somehow right themselves?

Standing here on the fabled Plymouth Hoe overlooking the sound, this strikes me as a profoundly moral question. Can humans, knowing that human civilization itself is at stake and that the outcome is uncertain, choose to turn a blind eye?

I turn to Young for answers. He writes—quite coolly—about what sets humans apart from other animals. He compares the human brain to that of apes. The brains are arranged in a similar way, except that the occipital and frontal lobes are better developed in humans. The occipital lobes give us the gifts of sight and visualization—the ability to see in a physical sense. Yet even more important to our survival, Young argues, are the highly developed frontal lobes:

> The frontal lobes, so far as is known, serve to maintain the balance between caution or restraint and sustained active pursuit of distant ends, which, above all else, ensures human survival in such a variety of situations, and makes possible the social life and use of language by which so great a population is maintained.

Connected to these frontal lobes, of course, is the quintessential human power of abstract thought—the ability to see metaphorically. To understand. It is, Young asserts, what makes us human.

When I go back to the question of whether we should just sit back and let the universe unfold, I have to conclude that it is not fully human to do so. To be fully human is to exercise restraint while pursuing far-off goals, to imagine that the future has the capacity to be worse than today, to recognize the signs that things are changing and not in a linear way. We are creating change on an unprecedented scale. That much is incontrovertible. The unknown is what future we will choose to create.

In all of humanity's 150,000-year history, it seems to me that this is the moment to harness the human gift of being able to plan.

*

The National Marine Aquarium in Plymouth was one of the first public aquariums in the world and the first in Britain. The original building, opened in 1888, comprised tanks in picture frames, to be seen from dark corridors. This new building, with its soaring ceilings, broad vistas, and inventive displays set right on the shore, opened in 1998.

Kelvin Boot, the aquarium's director when I visited, has a wit both quick and sharp. His sense of humor is dry. He is one of the few nonscientists I interview at Plymouth, and we bond immediately. By a pleasing irony, the aquarium is next to Plymouth's bustling fish market. "It's a wonderful juxtaposition," Boot says, laughing and pointing to the market. "You can buy them by the kilo over there. Here we spend money to keep them alive."

The fish market next door actually provides some valuable research data. Aquarium scientists nip over every now and again to see what's for sale. It helps them track changes in what's being caught in local waters.

As Boot explains, the changes are legion. When the new building opened, its centerpiece was the Mediterranean tank. It's the deepest tank in Britain, a soaring column of water teeming with creatures. You can stand under a thick, concave sheet of acrylic that reaches from the floor to far above your head and watch fish from another part of the world swim above you.

Now, however, these fish are not exactly from another part of the world. Today, these fish also live wild in British waters. The ocean off the coast of Devon has warmed up enough in just a decade that fish once caught in Provence are now common this far north.

A few years ago, Boot tells me, some of the scientists found some unusual, very young fish nearby. So the aquarium raised them, only to discover that they were Couch's sea bream, a species that had not reproduced in English waters in living memory. The classic Mediterranean triggerfish is breeding in British estuaries now, almost to the point of becoming a nuisance. Mediterranean sharks have been spotted here and there, and barracuda, more commonly residents of tropical and subtropical waters, now live in the ocean off Devon.

If the fish are moving, does it follow that plankton, the ocean's vagabonds, are too? To find the answer, I turn to the Marine Biological Association, the drafty home to so many Nobel Prize winners and to the Sir Alister Hardy Foundation for Ocean Science, known as SAHFOS. It is a world of wonder. While science in the world outside this building is moving at breakneck speed, at SAHFOS the technology for tracking plankton has remained unchanged for decades. Here are rooms containing handmade gunmetal bronze and brass contraptions—continuous plankton recorders—next to what looks like a watchmaker's repair shop where Roger Barnard mends the broken. There are no electronics. In fact, there's little in this place that you couldn't have found in the Camden Town of Charles Dickens.

For a flat £50 honorarium the recorders hitch rides with friendly ships, whose crews tow them through the waters of the world. Using the samples they collect, researchers can keep track of what plankton are where during what time. The prototype was a simple disk dragged by a wire behind a boat. When the disk came out of the water, the crew would eyeball it to see if it had caught any green gunk, meaning plankton. If there were plankton, that meant there would be fish. The idea then was not so much to find out about plankton as to map their presence or absence to increase the fish catch.

The devices are more sophisticated now, but just barely. These days they are rectangular boxes with tail fins that contain cloth woven from infinitesimally fine silk to catch the larger plankton. To preserve the catch, each piece of fabric, just one–four-thousandth of an inch thick, is rolled around a compartment packed with cotton batting that soaks up formalin from another compartment deep within the box. At one time, trying to be modern, some staff tried using nylon cloth instead of the fine silk. It was a disaster: too thick and hard to fold. Now SAHFOS supplies lengths of this fine, plankton-catching silk cloth to sister surveys in the United States and Tasmania.

If the machines haven't changed in more than seventy years, though, the goals of using them have changed profoundly. Now the idea is to examine the plankton themselves, to try to answer the globally critical question, Whither plankton?

Across the hall, in another dark and cramped room, scientists are examining rolls of silk that have come back from the sea, checking for about 200 types of phytoplankton, or tiny marine plants, and 300 types of zooplankton, or marine animals. Some of their custom-made microscopes, worth thousands of pounds, date back fifty years. Even a single-lensed, monocular microscope is still in use.

One of the scientists shows me what she's looking at. It is a little crustacean, a zooplankton called *Calanus helgolandicus* from the South Irish Sea. This fellow is distinct from *Calanus finmarchicus*, the type of plankton once rampant in these waters. She describes the differences between the two, trying to guide me as I peer through the microscope. *C. helgolandicus* is relatively large, has four swimming legs, and is concave. *C. finmarchicus*, with a fifth swimming leg, is more regularly shaped, smaller, and rounded.

I'm stumped. I can barely make out legs or curves or anything. The scientist chuckles. This particular plankton is folded over on itself, so it's a bit hard to see, she acknowledges. But, she says, you get used to telling one from the other.

Why does any of this matter?

C. helgolandicus, an important source of food for larval fish, has moved northward by an extraordinary 1,000 kilometers in the past forty years, replacing some of the *C. finmarchicus* that used to live here. But this new type of plankton is far less numerous than its predecessor—there's been a drop of about 70 percent in sheer biomass, or weight of living matter.

I go upstairs to the office of SAHFOS director Peter Burkill to see what he and his colleague Martin Edwards can tell me about the implications of these peculiar changes to the plankton. We have to talk above the call of the gulls on the sound and the wind whistling in the windows.

Edwards, the assistant director in charge of research, is one of the world's experts on the ocean cycles that determine such things as why one phytoplankton blooms just in time to feed the animals that need it. The field is known as phenology. As the climate changes and the ocean warms up, these finely tuned biological cycles are getting out of sync. It's called mismatch, and it's what happened to Kleypas's monkeys in Costa Rica; they died when the fruit didn't come.

One of Edwards's key findings, using research from the plankton recorders downstairs, has been that the mismatch in the North Sea already runs far beyond the tiny crustaceans his colleagues are looking at today. The dynamics of marine timing, he and a colleague wrote in the August 19, 2004, issue of *Nature*, "may have already been radically altered . . . and will continue to [change] in the coming decades if the climate continues to warm at its present rate."

For example, the little, ferociously abundant diatoms—marine algae with shells of silica—bloom in the spring just like plants on land. Traditionally, this has marked the beginning of a spring feeding and breeding frenzy in the ocean. The problem is that the trigger for diatoms' blooming comes from the length of the day, whereas other creatures are prompted to bloom or release eggs by the temperature of the water. So the diatoms have continued to bloom at the same time each year, even as the surface of the ocean has warmed. Meanwhile, the peak abundance for the larvae of such creatures as sea urchins and brittle stars moved forward a staggering forty-seven days between 1958 and 2002.

Edwards is fascinated with what these changes will mean for the fishery, especially when combined with chronic overfishing. But he's also looking at whether there are wider, so far unknown, implications for the plankton-driven cycles of oxygen, carbon, nitrogen, and silica. There will be consequences, he tells me, and they will probably be negative.

Burkill, who has been tossing in comments from time to time, springs to life at the mention of the big picture. This is where his heart is, in studies of how the changes to ocean and climate will affect the whole planetary system. He has made a career of studying even smaller plankton than the bronze recorders downstairs can capture, including the bacterial sort that help distribute critical minerals in the ocean's depths to the larger plankton that need them to grow. He starts to tell me a story about these tiniest plankton—one of those causal chains of events in this complex system of ocean and air that I could never have imagined.

During the 1990s he and a group of other scientists spent a great deal of time doing shipboard research at a hotspot in the Arabian Sea, at the top of the Indian Ocean. As concentrations of carbon dioxide and other greenhouse gases increase in the atmosphere and change the global climate, the powerful southwest monsoon blowing over the Arabian Sea becomes even stronger. Its winds churn the water more and bring extra chemical nutrients from the deep water toward the top. Just like the crop fertilizers that run off into the Gulf of Mexico, the glut of nutrients pushes the phytoplankton into an orgy of eating and reproducing. These marine plants die quickly and fall to the bottom. There the mass of plant bodies supplies a second feeding frenzy, among the sea's bacteria, which produce vast quantities of methane, nitrous oxide, and sometimes, as oxygen is depleted, hydrogen sulfide as waste products.

Already the Arabian Sea has far more than the usual oceanic concentration of methane, and some of that is flowing into the atmosphere. One of Burkill's papers calculated that the flow of methane from the Arabian Sea into the atmosphere is five times greater than the ocean's average, albeit still a tiny amount in the scheme of things. Most of the plankton-generated methane remains in the water because the surface and the deeps of the sea are separated into layers that are hard to mix. But over the past thirty years, the distance between the layers in parts of the Arabian Sea has become alarmingly narrower.

If the store of deep-sea methane should ever breach the surface—think of a nasty belch—it would release a tremendous amount of methane into the atmosphere. Methane is a far more potent greenhouse gas than carbon dioxide. Although it lasts for a shorter time in the atmosphere, a sudden large release could prove such an assault on the climate that it would represent a tipping point, the moment when the global system begins to work hard not to recover its old state but to create a new one.

As he talks, Burkill has been moving around this light-filled office, collecting bunches of scientific papers to give to me. He'd like to go back to the Arabian Sea to do more research there, he tells me as I leave. He wants to know what the plankton there are doing right now.

I'm staggering by the time I get back out onto Plymouth Hoe. It feels as though the more I learn about plankton, the more questions I have. As for the biggest question—whether any of these changes will affect the global ocean's ability to support life—that's also the least settled of all.

*

Humanity has tackled other complex issues, like discrimination based on skin color, gender, or sexual orientation. For society to be convinced that these were real, were problems, and needed to be made right, a whole bunch of dots needed to show up on a screen.

On the issue of racial discrimination in the United States, for example, a classic technique in the 1950s and 1960s was to look at collective numbers. Certain statistics were considered proxies for the vitality of the black community as a whole. Compared to that of the white population, what was the black infant mortality rate? The black life expectancy? The proportion of African Americans who owned property? What was the average black income compared to that of whites? How many African Americans were in positions of corporate or political power? For each of these individual indicators there were exceptions to the rule. But when you connected the dots, they told a story.

The story of plankton can be approached in a similar way. It's hellishly complicated, but a few key points are clear: that plankton are extraordinarily important to the workings of the planet, that they are changing dramatically in a very short time, and that the potential implications of this are severe. It's also clear that some of the best minds in the world are working feverishly to figure out the implications. Plankton are telling us a story that we ignore at our peril.

By the time I get back to the Plymouth Marine Laboratory building to the office of Manuel Barange, director of the Global Ocean Ecosystem Dynamics program (GLOBEC) all I really want is reassurance. Luckily Barange, trained as a fisheries scientist, is a philosopher at heart.

GLOBEC is a muscular international group devoted to answering precisely the sort of questions I've been asking. When it was set up about two decades ago, Barange explains, its goal was to find out what was happening below the fish. Even in 1999, when GLOBEC became part of the blue-chip International Geosphere-Biosphere Programme, the understanding of global change was "still ropey," meaning that it wasn't the accepted wisdom. Today the GLOBEC program is all big picture, all the time.

It's incredibly hard, Barange says. "This problem is more complex than we ever imagined. Every time you solve a little bit of the puzzle, you see that there's another huge part of the puzzle unsolved over there."

For example, his team is trying to develop a vulnerability index that will predict changes to the productivity or metabolism of parts of the ocean. Key to the whole thing, says Barange, is the need to deal with the climate issues in conjunction with human issues. In other words, humans are now an intrinsic part of the marine system.

"It's not one without the other," he says. "It's never humans and not climate. It's always both. This is a frontier we have to break. We need to break disciplines."

We talk for another hour or so and I start to see glimmers of hope. Near the end, he leaves me with this: "The scale of the solution has to be to the scale of the problem."

It is the scariest thing I have heard yet.

Reading the vital signs: Fecundity 5

PANAMA — The blind wisdom of the coral

Being a successful coral means having a superb sense of timing. And not the arbitrary, human-built timing that measures from past to future, but the primeval sort that counts off rhythms. In other words, not progress but pattern. Not the measurement of something fleeting but of a constant.

Spawning corals survive because they understand constants, mysterious signals that come once a year from an alignment of sun, earth, and moon. All year these primitive animals, which have neither brains nor eyes nor the ability to move, laboriously store energy to prepare eggs and sperm for the spawn. Then on one preordained night, between five and seven days after the harvest moon, at least for some coral species, billions of them send forth all this genetic material into the wide ocean in an ancient pageant designed to make baby corals. It's like a ritual, a once-a-year orgy of group sex.

Humans figured out only in the mid-1980s—during night dives on Australia's Great Barrier Reef—that this is how corals reproduce. But ancestors of modern corals have been found in fossil records going back 450 million years, and it's a safe bet that some version of this annual rite has been going on for about that long.

Corals sense movements in the universe that humans appear to have forgotten. Viewed through the lens of nonhuman time, they seem to be the ultimate survivors. More than survivors, though, they are progenitors. The bony reefs the corals build are the nurseries of the ocean, home to at least a quarter of all known life forms in the sea. Because the planet has far more ocean than land, and because the ocean holds so much life, coral reefs are biologically among the most critical parts of the planet. Yet the reefs are one of the places where all the important human-caused threats to the seas intersect. The biggest of these is global climate change, which is affecting ocean temperature, volume, acidity, and possibly salinity, and maybe even the structure of the currents, in ways scientists are just starting to calculate. Moreover, overfishing and pollution are harming the ecology of the reefs.

The result is that 20 percent of world's reefs have already been destroyed and another 50 percent are in trouble. In the Caribbean, 80 percent have died in the past three decades. So here in the Caribbean, if the corals are still reproducing, in perishingly hot waters, that will be a sign that the ocean still has some moxie, that it has not fallen into reproductive unconsciousness.

These primitive marine animals have more to teach us. What would we see if we could view the planet not through the linear time of human stories but through the multidimensional time of corals and their ilk? Would we see more clearly, recognize the great dangers we are running? Could we, too, become survivors in the long game of evolution?

I have decided that I need to watch the corals as they spawn in response to the lunar cycle, to look for life in the larger theme of death. And this is why I am arriving in Panama on the night of the full moon nearest the autumnal equinox, day zero in the lead-up to the spawning of the corals.

Nancy Knowlton is the corals' timekeeper. A superstar marine biologist who is now the first holder of the Sant Chair in Marine Science at the Smithsonian's National Museum of Natural History in Washington, DC, she has also worked extensively with the Smithsonian Tropical Research Institute in Panama City and with the Scripps Institution of Oceanography in La Jolla, California. Tall and rangy, with a thick fringe of brown hair covering her forehead, Knowlton is best known for showing scientists that the oceans hold far more species than anyone had thought, probably by a factor of ten. She's also renowned for being able to identify marine animals that are difficult to tell apart by conventional methods, the so-called cryptic species of the seas.

For instance, she was the first to realize that the corals we're here to monitor—*Montastraea franksi*, *M. annularis*, and *M. faveolata*—are three species instead of one. These three are the slow-growing workhorses of the Caribbean reefs, with boulder-shaped bony structures that can get as big as a house. Understanding the future of the savaged Caribbean reefs means knowing these *Montastraea* inside and out.

For some reason she can't fully explain, Knowlton loves corals. Perhaps it's the tension between how primitive they are and how ecologically critical. Or the sheer intellectual pleasure of trying to understand an animal that early generations of scientists thought was a plant. Even now, most scientists don't give corals much thought.

The love affair started in 1980 when Hurricane Allen, one of the strongest hurricanes in recorded history, hit the Jamaican reefs Knowlton was studying. Four days later, before the storm had fully subsided, Knowlton was underwater at the Dancing Lady Reef in Jamaica's Discovery Bay to catalog the damage.

The reef was pulverized. Knowlton figured she would study it for years to come and chronicle its recovery. But it never came back. It just kept dying, hit by disease and hindered by an ecosystem that had already been badly damaged by overfishing. It was the beginning of the end of the Jamaican reefs, the textbook case of a biological worst-case scenario.

Once Knowlton worked out that the lack of fish and the health of the corals were connected, she had to keep going. She's become one of just a small group of people in the world who understand corals. For one thing, she tells me, the hard bony bits that we terrestrials think of as corals—the pieces we put on our mantles and wear as jewelry—are the inanimate parts of the animals, the calcified base the coral animals continually add to. As Knowlton puts it, it's as if we grew bones out of the bottoms of our feet and kept getting taller. Or like a tree trunk that grows taller throughout its life so its leaves can reach for the sun. Except these bony pieces are stone, created by the coral from the carbon and calcium in the ocean through a simple chemical process.

The shape of these bony pieces distinguishes some corals from others, but that's not precise enough to tell species apart. It's also key to look at the living animal: the flaccid layer perhaps 100 cells deep that covers just the upper surface of the bony structure. This has been one of the great obstacles to identifying corals and many other marine animals. Generations of taxonomists have looked at them on land, dead. But they look far different alive than they do dried or frozen.

Knowlton decided early on that if she wanted to study marine animals, she needed to go where they were, become part of their environment, no matter how foreign it felt to her. It means she has spent a great deal of her life underwater, attached to a scuba tank. Now, it's second nature to her. She regularly sees creatures that most humans cannot imagine exist.

She loves Panama too, despite having been taken hostage from a Smithsonian research station by several Panamanians when the United States Army invaded the country in December 1989. Knowlton and her daughter, along with nearly a dozen Smithsonian employees, were forced to walk barefoot across hills for most of a day, kept without food for two, and finally abandoned by their captors. A U.S. Army helicopter pilot found them.

Knowlton doesn't dwell on that. At one point in our journey together she says she'll tell me the whole story, but we never quite get to it. The closest we come is when she shows me a photograph of the group of hostages taken after they'd been whisked back to the American army base. It's pinned up beside her desk at the Smithsonian research base, along with drawings from her daughter's early school years. In the picture Knowlton's smile is huge, relieved. Her daughter, a child then, stands close.

Knowlton talks of the grace and friendliness of the Panamanian people, the ferocious beauty of the rainstorms that she watches with glee from the balcony of her flat in Panama City, the engineering feat that led to the building of the Panama Canal. Now she is based in the United States, and her flat in Panama City has become a refuge for graduate students in marine biology and for itinerant academics. And authors.

Knowlton and I arrive at her flat to stay for a couple of nights before leaving for the research station at Bocas del Toro on the Caribbean side of Panama. There, we hope, the corals will be spawning in about six days.

Like high tide, dawn, and sunset, a full moon has a specific time. That time changes from month to month, determined by the precise moment when the sun and moon are on opposite sides of the earth. This month it happened on September 18 at 2:01 a.m. Universal Time (also known as Greenwich Mean Time). According to Knowlton's painstaking years of study and calculations, this means that the corals will spawn about six days from now, around September 24. The *M. franksi* will let loose their slick of genetic material 100 minutes after sunset; *M. annularis* and *M. faveolata* will do so 100 minutes later.

It's not clear why this happens after this particular full moon. Again, it's one of those cues that the corals recognize and humans don't. One theory is that the corals respond to particular intervals between sunset and moonrise at this time of year.

*

It is shortly after dawn on day two, and Knowlton is anxious. I can tell because her foot is in constant movement—up and down, up and down. We are on a Jetstream 31 flying from Panama City to the rickety resort town of Bocas del Toro in the north of Panama. The Smithsonian has had a marine research station here for several years, on a prime piece of Caribbean waterfront. Originally a sweltering shack, the station has become more sophisticated in recent years. Not only are there modern, air-conditioned labs filled with pristine equipment but also a decorative man-made pond, filled with caimans, that winds around the buildings, and landscaped grounds complete with hanging poisonous snakes that the gardener periodically chases away.

About a dozen students and scientists from all over the United States, and two from Taiwan, have assembled at the behest of the corals, with hundreds of kilograms of diving gear and lab materials, to help Knowlton figure out whether these enigmatic animals are successfully reproducing. The plan is to scuba dive onto the reef for hours at a time, chronicle exactly when the corals pop their tiny pink bundles of sperm and eggs, follow the bundles to the surface of the sea, collect some of the spawn, and then stay up most of the night checking the eggs under a three-dimensional microscope to see how many get fertilized. It's an expensive and time-consuming enterprise. But Knowlton's team is primed and in place. This is not what she's worried about.

Her concern is the corals. Although she and her team have correctly predicted the timing of the mass spawn for years, there's always the chance that the corals won't do what everyone is expecting. Knowlton might have mistimed the event, which would be an expensive mistake, given the cost of bringing in all this high-powered expertise. There's an off chance that the corals spawned last month, or that this year they will go for broke a few days early. Or it could be that her timekeeping is correct but the corals are too sick and listless to spawn.

Aboard the airplane, Knowlton has slipped off her shoe. Her bare foot is moving nonstop. She is wondering what the corals are doing. Are they making their bundles of sperm and eggs? Are they responding to the hidden cues from the moon, earth, and sun? What if the corals don't spawn? What would that say about the future?

As soon as we arrive at the research station, we get bad news. Davey Kline, a coral reef ecologist who was one of Knowlton's students and who lives in Panama, has been diving in the reefs all around this section of the Caribbean. The corals are badly bleached. In fact, it's as bad an episode as he's ever seen. Bleaching is one of the great modern scourges of coral. It means that the symbiotic relationship between the coral animal and the algae that live within its cells is breaking down. The algae, which photosynthesize food for the corals and also give the corals their jewellike colors, die, leaving behind white coral tissue. The corals starve unless they can attract more algae.

I ask what causes the bleaching.

Knowlton doesn't hesitate: the water is too warm for the algae to live.

Why is the water warm?

Again, the answer comes right away: it's global climate change, the heating up of the atmosphere—and, therefore, the global ocean—due to high concentrations of greenhouse gases. Both the frequency and the intensity of coral bleaching have been increasing steadily since the 1980s, in lockstep with ever higher concentrations of these gases.

There is one bright spot, though. Kline notes that Knowlton's three species of *Montastraea* don't seem to be as badly bleached as some of the 100 or so other species of corals in the Caribbean. Knowlton breaks out into huge smile. The team laughs in relief.

"They've put up with our lights and our chisels for so long, the temperature doesn't bother them," she says. She knows, of course, that it's something different. Maybe the specific type of algae that live in the *Montastraea* can take higher temperatures than the algae in other corals. Maybe it's something no one has thought of yet. Another coral mystery to work on.

By 10:45 a.m. the air-conditioned lab where Knowlton's coral experiments will take place has been fully kitted out. I begin to think of it as the war room, the strategic center of this minutely planned operation. Students and scientists have set up cupboard after cupboard of carefully labeled silicon tubes, Petri dishes, pipettes, vials, culture trays, centrifuge tubes, hammers, chisels, buoys and lines, flashlights, batteries, knives, wire brushes, hundreds of meters of bright orange flagging tape, and dozens upon dozens of glow sticks. High-powered laptops are plugged in on several of the counters, blasting music and spitting out data. Knowlton is standing in front of a large white board, marker in hand, making meticulous lists.

It's only when we embark on our reconnaissance mission in the afternoon to find the transect of corals that Knowlton has been tracking for four years that the extent of the bleaching becomes clear. Knowlton points over the side of the boat to show me. I see what she sees: an expanse of neon white under the boat. Even through several meters of water, it looks alien.

One of the students, Nikki Fogarty—who is trying to make a career of studying corals—shakes her head. It's depressing, she says. She's not sure whether there will be enough corals to study by the time she's finished graduate work. She's been scuba diving for the past fifteen years, since she was just a kid, and has seen the mounting destruction close up. Now every coral reef symposium she attends shows charts tracking global declines in the number of healthy reefs. At those symposia, as on this big boat, there's a question no one wants to deal with: is all of this research too little too late?

Soon, though, the joy of being among the corals takes over. Knowlton and nearly a dozen others will be underwater for hours, right down to the final measures of compressed air. When all the scuba tanks on board are empty, some of them start snorkeling.

Because this bit of the reef has been studied for so long, many of the coral colonies have been marked with tags. But since the last spawning a year ago, the tags have become encrusted with ocean life, taken back by the sea. Today the crew has to laboriously find each tag, scrape it so the number is visible and make a notation of where it is. This is critical to the record keeping of the whole enterprise. And it's exhausting. But it's the only way the team can keep track of whether an individual coral spawns and compare it to what happened in other years.

I'm still on the boat. It's a huge, sweltering creature, affectionately dubbed "the floating classroom," with room for eight plus their diving gear and extra tanks. I realize after everyone else has jumped in that I can't make myself go under the surface of the sea. It must be primal, this fear, residue of a species whose distant ancestors abandoned water in favor of land. Some people can overcome it, can flourish in the ocean. I can think only of death.

The deeps seem unknowable. Throughout human history we have seen the surface of the sea, imagining and fearing what's below. Down the ages, through legend and myth, we have made the oceans the home of monsters, shape-shifters, demons, and danger. Here are the gaping teeth that make us shudder, the whirlpools that suck us to places beyond redemption, the predators we can neither hear nor smell.

It's the flat two dimensions of the surface we've tried to master: the trade routes; the best fishing spots; the arts of avoiding storms, shoals, and icebergs. For very good swimmers, the known sea extended perhaps 30 meters down, as far as the breath could hold.

My urge is to go so much deeper—both physically and psychologically—to know what it's like to be part of the sea in all of its dimensions. And then to understand them across the long fourth dimension of time—

from past to present to future. If you put all four dimensions together you get an integrated whole, a coral-style understanding of life in which things happen in patterns, in cycles, in ancient rhythms not easily visible. Putting the four dimensions together is what the Australian Aborigines think of as Dreamtime, or anywhen.

In the span of human history, only a few of our species have had the ability to consider even the third dimension of the ocean—its vast depth—much less the three dimensions over time. It was only in the 1920s and 1930s that humans figured out how to compress air and send people as far down as 150 meters. And it was the 1970s before deep-diving submersibles routinely got as far down as about 1,500 meters. And that's still a small fraction of the real depth of the ocean—which is on average 3,794 meters deep and at its deepest 11,034 meters.

Of course, it's not just about depth. It's also about volume, the three dimensions together.

Life on land and in the air occupies a thin layer above the surface. Most of land life is right on the surface, with the exception of a few high-flying birds and the odd skyscraper. It's a narrow third dimension, in other words.

In the ocean, this dimension is thick. Life runs across the surface in all directions and down to the bottom. The dimensions move and connect on a scale that land dwellers can barely fathom. In fact, when you add up the earth's biosphere—the part of it that is available for living creatures—the land portion comes out to just 1 percent of the total volume.

Not only that, but if you look at the phylum level of the tree of life (the second-highest level of taxonomic organization, beneath kingdom), it's astonishingly sea-based. Of the thirty-three phyla of animals on earth, twenty-eight live in the ocean and only eleven on land. (There's some overlap.) Thirteen evolved in the ocean and are still found only there. Just one, the ancient, sluglike Onychophora, evolved on and remains restricted to land. Not only that, but animals in the ocean—not hampered by the land's gravity—boast twice as many body blueprints as animals on land. Theirs is a far more diverse and rich living space.

*

Tuesday, day three in the buildup from full moon to spawn, is a day of victories. Knowlton and eight members of her crew are in the boat by 12:15 p.m., chisels and magnifying glasses in hand. The water is clear, the day sunny. We can see easily to the bottom of the transect; corals loom under the boat. Today, I am determined to see them close up.

The others plunge in, attached to tanks. I resolutely press my mask and snorkel to my face, put on borrowed flippers, roll over the side. My nose fills with water—the mask doesn't fit. But I can see corals and the whole marked section of the reef where the scientific studies will take place. I'm hanging close to the boat, just above Allen Chen, a marine biologist from Taiwan who is here to see the action, and one of his graduate students. They are moving slowly, deliberately. This is not a business for show-offs.

The water is full of moon jellyfish today. They sting, but not like some of the more dangerous jellies, such as the man-o'-war. Knowlton has given me the spiel on these fellows. Before she went under, I asked her—attempting to be matter-of-fact—what can kill me out here. She pondered this a minute.

"You know about the man-o'-wars?" she asked.

"Well, actually, no."

"They have tentacles hanging down. If you got them wrapped around your face, near your nervous system, that would be a problem. Try to avoid them," she advised.

Hanging on the surface, moving with the waves, checking for bundles of tentacles, I watch the moon jellies warming themselves in the rays of the afternoon sun. The sun comes down in shafts, refracted through the waves.

I see brain coral, so named because it looks like a human brain. And loads of *M. franksi*, Knowlton's beloved species. They're frankly hard to bond with. Unmoving. Seemingly dead. These are not the warm and fuzzy mammals that most conservation organizations put to the fore to convince humans that life is valuable. I'm thinking about those cute pandas and adorable baby seals. Even lemurs. These corals could be so many rocks strewn on the shallow ocean floor.

A while later, Knowlton bounces up. She's been down on the reef with her chisel, whacking off tiny bits of coral to see whether they're "ripe," meaning, whether they're making eggs and sperm for the spawning. She's grinning, tension dissolved. The timing is correct. The spawn is going to happen.

Knowlton has also brought up buckets of sea creatures from the reef for me to look at under the three-dimensional microscope. Once we get back to land, we take some of these marine goodies up to a lab. She puts a tiny piece of reef, covered by seawater, under the microscope. It's solid with organisms. At least twenty-five, at a conservative count. Here are the bizarre bryozoans, who make stacks of clear glassy boxes. Almost turrets. Their feeding apparatus emerges from a hole at the top of the stacks. There's a barnacle, its feathery appendages coming out the top. Here's a clam too small to see with the naked eye.

A brittle star, the smaller and more flexible cousin of a starfish. Several types of sponge, some bright orange and others chartreuse. A sea anemone, which Knowlton thinks of as a coral without a skeleton. It's hungry and has captured a minuscule shrimp. And over here, a worm. This minute piece of reef is swarming with life.

Knowlton is in her element, finding these creatures, handling them with utmost gentleness. She must have spent hundreds of hours looking at similar batches of life under a microscope, yet she's as absorbed in this one as if it were her first.

The question hangs over us: if there's this much diversity of life in a tiny piece of one type of reef, how much is there in a whole reef ecosystem? Putting a number on what's there and what's being lost will be Knowlton's defining work, the thing no one has done before.

*

Day four. Great sorrow. Fogarty and a helper, Katie Lotterhos, went out last night to capture some eggs and sperm from a type of coral that is supposed to spawn a day or two before the *Montastraea*. They are looking at the elkhorn corals, *Acropora palmata* and *A. cervicornis*, which are heavily endangered in the Caribbean and elsewhere. In the last several years 80 to 90 percent of the elkhorns in the Caribbean have died.

Fogarty and Lotterhos were trying to get enough elkhorn sperm and eggs to cross the two species in the lab and then grow them to find out whether the hybrid offspring would thrive. They spent a full day taping off a countertop in the lab into a matrix and labeling each square, painstakingly measuring distilled seawater in pipettes. Then went out, in high spirits, to capture the spawn. Then . . . nothing. No spawn. No evidence of spawn. The human calculators of time have been skunked.

Knowlton holds back from comment. She knows it's devilishly difficult to predict when corals will spit out their bundles of eggs and sperm, and she's not ready to criticize graduate students who have mistimed things. Still, she knows that her *Montastraea* are likely to spawn on day six, two days from now, and that means we go through a full-scale operation tonight, as if today were the day. She's not about to miss anything if she can help it.

I will have to do my first night snorkel tonight. Knowlton gives me my day's assignment: by dusk, she wants me to be able to dive to 3 meters with a snorkel. It's a very different skill from swimming placidly along the surface, as I have been doing in seas all over the world. She assigns one of her graduate students, Kristen Marhaver, a brilliant geneticist, to teach me. Marhaver, whose name means "keeper of the harbor" in German, reminds me of a mermaid. Long, slender, with a shock of red hair, she seems more at ease in the water than out of it.

We take the boat and get on our gear at a shallow reef neither of us has seen before. Marhaver does an elegant five-minute lesson on diving with a "snork" while I try to keep up. Then she starts with direct incentives. She dives to the bottom to look at what's there, challenging me to follow. She loves every creature. Every molecule. She can't keep her hands off them. She keeps surfacing, enchanted, with new things to show me. Her knowledge of reefs is encyclopedic.

One quick look around and she sees that there are far too many algae. In the good reef/bad reef rubric of Katharina Fabricius of the Australian Institute of Marine Science, this would be a bad reef. A healthy reef, Marhaver explains, would have loads of sea urchins to eat these algae. In fact, there used to be so many sea urchins in the Caribbean that scientists considered them a nuisance. Then disease wiped them out.

With the sea urchins gone, the algae take over. Two specific types have flourished, Marhaver tells me. One is bitter. Fish fed pellets containing this type of algae will spit them out. The second is filled with calcium carbonate—the same stuff the corals make their bony bits out of. It's too crunchy for any sea creatures but the urchins to eat. So the algae keep growing.

It's terrible for corals. Baby corals need hard surfaces to settle on; these algae keep everything slick.

The fish are absent, fished out by locals. We see one crab. No lobsters. No sharks. No baby corals, just the older, established ones, smothered with algae. It's a wasteland.

Marhaver puts on her scuba tank and leaves me to play. Periodically she surfaces to point something out. I try and try to descend, hold my breath, kick as hard as I can. Finally I decide I need some weights around my waist. I am, as Knowlton delicately puts it, rather buoyant. Meaning I have a nice solid layer of fat that keeps me on the surface, like an inflated inner tube.

It's a bit crushing. But not as bad as when I try to hoist myself into the boat. I've forgotten to let down the ladder beside the engine; it's held tight with stretchy cords. Everyone else is underwater, probably for another hour or two. I keep thinking about man-o'-wars. About whether there's a shark in the Caribbean that hasn't been killed off and might like a buoyant morsel of human for lunch.

Hanging in the water near the boat, I consider a book I found in the library here at the research station—*Air and Water: The Biology and Physics of Life's Media* by Mark W. Denny. It describes the physical differences between air and ocean. These different media help determine how different creatures are shaped, how they hear and see. Water, for example, is a far better conductor of electricity than air, by a factor of twenty billion.

One of the fascinating, relatively unexplored aspects of life in the ocean is how marine creatures communicate through electric currents and how they might pick up electrical signals from deep within the earth's core. Denny calls this a "sixth sense." Every time a muscle contracts—say, when an animal breathes or tries to heave herself over the side of a classroom boat without a ladder—the body sends off electrical signals that could be read by other animals.

This could be a clue to the blind wisdom of the corals. They certainly respond to light, to pressure, to chemical signals from other creatures. Could they respond somehow to electricity in the water?

I'm lost in these thoughts until I remember that Denny says sharks and rays are among the most alert to electrical impulses. I decide to remain very still.

Marhaver shows up unexpectedly. She's come back to check on me. I explain that I can't get back in the boat without a ladder. She looks at me quizzically and slips effortlessly over the boat's side to let down the ladder before popping back down into the sea.

By 6:53 p.m. it's pitch black and we're back out at the transect, in the midst of what I can only think of as a military operation, to witness the corals in the act of the spawn. No one believes that today is the day, but we're acting as though it is for practice purposes.

Don Levitan, a wiry marine biologist from Florida State University and Knowlton's research partner on the coral spawning, has arrived with three more students. It's now a team of more than a dozen highly educated scientists looking for coral spawn in the waters of the Caribbean.

Underwater the mapped section is lit up with chemical glow sticks.

Each pair of divers is assigned to one point on the transect with the task of examining each of hundreds of coral polyps for a total of nearly three hours, with one 10-minute break in the middle to change compressed air tanks. I'll be snorkeling on top of the water.

We're all waiting for the corals. The divers are munching on chocolate bars for a last-minute burst of energy.

Knowlton sees the worms first. Perfect circles of bright white light on the surface of the sea around our boat. They are female glowworms, trying to attract males. In a few minutes we see luminescent puffs beneath them. These are the males, drawn by the light, spraying the females with semen. Knowlton grins. The glowworms always mate one or two days before the corals spawn.

By 8 p.m. the divers are in, eyes glued to the corals, watching for signs of the bundles of reproductive material. Each individual coral of these three *Montastraea* species is a hermaphrodite and produces both sperm and eggs, bound together chemically. A few minutes before the corals release these bundles, the bundles "set." That means they rise through the flesh of the corals and sit at the single orifice, ready to burst at a signal only they can discern. Buoyed by the fat in the eggs, the bundles rise to the surface, then break up into just sperm and just eggs.

Marhaver has explained to me this brilliant mechanism for collapsing the three dimensions of the ocean world—plus the fourth of time—into just two. Instead of sending their genetic material into the wide, three-dimensional ocean at random times, hoping the short-lived eggs will meet up with sperm from the same species, the corals send their bundles to the surface, which eliminates the dimension of depth. And they do it at the same moment, which eliminates the dimension of time. If some of the corals release their eggs and sperm at the wrong time or on the wrong day, they don't get to pass on their genetic material. It's a pretty smart strategy for an animal that can't walk or swim around to find a mate.

I wave my hands in front of me in the water. The plankton light up, bioluminescent. It's magical.

Then fear swallows me up. I remember all that I don't know, can't yet understand about this water of life. Is there anything with long tentacles hanging near the surface? I sweep my light around, then remember that marine animals are attracted to light and switch it off. My mask fogs up. I'm not wearing my glasses—how far can I actually see? And I know that anything that could hurt me can swim a heck of a lot faster than I can.

I'm ticking off the minutes. The divers below me are crowded around the transect, peering at the minute orifices of each coral polyp to see if they're ready. All at once, this seems slightly insane. And in the end, nothing. The corals are not spawning.

Still, there are hopeful signs that something will happen in the next few days. The reef seems to know when the corals are about to spawn, and other creatures spawn in response. And tonight they are sex crazy. The little brittle stars are standing up on tiptoes, arching their backs, and spraying each other with sperm and big ruby-red eggs.

*

On day five, only four corals spawn. No eggs are fertilized. Everyone is exhausted.

*

Day six. From the second Knowlton gets into the water, she knows that tonight is the night. The water is electric. Even land creatures like us can feel the sexual charge. It's as though the whole reef is vibrating.

It's 7:40 p.m. The first group of divers is already underwater, scouring the reef with high-powered flashlights, prepped with red glow sticks to mark anything interesting. At 7:43 Knowlton spots the first colony of corals getting ready to spawn.

Then, right on cue, it starts. The first bundle of eggs and sperm that the scientists are tracking heads up at 8:05, then another at 8:06, 8:08, two at 8:11, one at 8:12, 8:13, 8:14, 8:18, and then a couple of laggards at 8:21.

Divers follow them up, releasing red glow sticks to follow the bundles. This enables the two members of the team who are in a rowboat to follow the eggs and capture some of them in plankton nets. I'm hanging out on the surface again with my mask and snorkel. The air fills with a pungent salt funk. The whole reef is having sex. I can't see the bundles pop out but I can see them rise, pink and delicate, in the arc of my flashlight. And I can see the tiny creatures who are gorging on all this great egg and sperm protein. It's a feeding frenzy as much as a sexual one. This is what all the reef organisms have been waiting for all year. It is a moment of exquisite hope.

Suddenly, though, there's lethargy. As if the reef is having a leisurely smoke at the end of it all.

The divers surface to change spent tanks and plunge again. By 9:20 the other two coral species are setting their bundles. At 9:46 the second spawn has begun, and the surface again fills with ectoplasm, the salt smell of sex, and the eerie red light of the scientists' chemical glow sticks.

Finally, the spawn is done. Triumphantly we take our night's work to the lab, calling ourselves Team Spawn. We've gathered coral eggs. Now we are eager to see what percentage have been fertilized and will start dividing into coral embryos.

By 4:15 a.m. hope is beginning to fade. Some years, every single egg sampled has been fertilized; tonight it's about half. Plus, team members—waterlogged and bleary-eyed—have added up the numbers and found that only about half the coral colonies that could have spawned did.

Biologically, these are not great odds. This ancient, magnificent rite of nature, so painstakingly cataloged tonight, has probably failed to produce enough baby corals. On the other hand, in the whole long history of coral spawnings going back perhaps 450 million years, humans have witnessed and tried to catalog just a few. Maybe in the grand scheme of things—in coral time—this is a pretty good spawn, given the bleaching and other exigencies.

In any case, all of the biologists gathered here at this research station know that the problems facing the corals are larger than whether this particular spawn was successful. They know that corals all over the Caribbean—and in many other seas—are in dire straits and declining in abundance. Scientists rarely see baby corals in the Caribbean any more, partly because the coral embryos that do make it through the spawn have trouble finding a clear spot on the seafloor to attach to. Without that, the embryos die.

They know that as the sea becomes more acidic, the delicate coral embryos might not survive and that those that do will find less calcium available to make their bony mountains. They know that the corals are jeopardized by the pollution that pours from fields and coastal towns into the shallow lagoons the corals need to survive, and by rising water temperature, which kills the microalgae that make the corals' food.

Where will it end? And how? Can the corals recover? What does it mean if they don't? When Knowlton became a marine biologist, it was for the thrill of understanding these extraordinary creatures. For Marhaver, Kline, and Fogarty, it's a race to save them.

Down the hall, Dan Distell is preparing for the worst. For days now he's been following Team Spawn like a wraith, methodically collecting samples of the corals' genetic material for the Ocean Genome Legacy, a nonprofit organization based in Massachusetts. It's a small hedge against the corals' lack of fecundity, a biological contingency plan aimed at preserving genetic building blocks of marine life in case the animals and plants themselves stop being able to produce them. Distell has already confided to me that corals are a priority of the Ocean Genome Legacy because they are the most urgently endangered group of species known of in the world.

Reading the vital signs: Life force 6

HALIFAX, CANADA

The fate of the fish

Sometimes, in order to discern the present, you have to be able to see both the future and the past. Time must be compressed into a single dimension. What you are able to see at any one moment may be indistinct, like a blurred or badly exposed photograph, but a series of many exposures layered on top of each other can sometimes provide a clearer picture. If you lose the past, the present and the future cannot really be understood.

So it is with fish. If you don't know what used to be there, how can you judge what's there now or what that implies over time?

The condition of fish is one critical, obvious indicator of the health of the ocean, particularly since fish have been an integral part of the human food chain—and human consciousness—for thousands of years, leaving their traces in art, cookbooks, literature, and garbage heaps. Until this generation of scientists, however, hardly anyone thought to investigate fish-related trends, either in a specific part of the ocean or across the global ocean itself. This is a brave new field.

One of the most exciting universities in the world for this new field is Dalhousie in Halifax, on Canada's east coast. Here, marine biologists cut their teeth in the 1980s and 1990s on the legendary collapse of the Atlantic cod in the sea outside their door. They read the vital signs of that population, delivered the prognosis that the cod were in trouble from overfishing, and were roundly ignored until the cod did indeed effectively vanish as a commercial and biological entity in that part of the sea. Even though cod fishing has been banned in that region for more than fifteen years, the cod have failed to come back. It's not clear if or when they ever will. This biological conundrum has spawned much of the drive at Dalhousie to read the future of other fish.

Will other fish go the way of the cod? How will all the other changes in the global ocean—rising temperature, rising acidity, falling metabolism, dropping fecundity—interact with overfishing, and overfishing with them? What effect will they have on the ocean's overall capacity to support life? Could the global ocean ever really lose its ability to keep fish alive? And if it did, how would that affect the other creatures that live in the sea, not to mention the billions of humans who rely on fish for food?

Three Dalhousie scientists are among those on the cutting edge of figuring out the future of fish, and I've spoken to each of them many times before. Their work is part of the global Census of Marine Life, which is trying to map what's in the ocean, what used to be in the ocean, and what will be in the ocean over time.

Heike Lotze is an expert in reconstructing past marine and shore systems, in seeing what's no longer there. She grew up in the north of Germany, a few kilometers from the Wadden Sea, and has pioneered this type of biological detective work. Her work is fascinating because she relies not only on archaeological findings, which tell of life thousands of years ago, but on poetry, medieval maps, art, and prose to discern the types of marine animals and plants that were once there.

Boris Worm, who is married to Lotze, goes in precisely the opposite direction. He is a soothsayer, using complex mathematics and computer analysis to calculate present and predict future populations of fish. The ecological forecasts that he and Ransom Myers, the third Dalhousie scientist, have published over the past few years have startled the world. Most notably, in 2003 they published a paper in *Nature* explaining that the populations of all the big predator fish in the ocean have plummeted by 90 percent in the five decades or so since modern industrial fishing took hold. And the fish that remain don't grow nearly as large as they once did.

This paper—known as the "90 percent paper"—was widely publicized. It forms the basis for the widespread sense that something, somewhere, is terribly wrong with the fishery. It's hit the public discourse far more powerfully than the problems with pH, oxygen, or plankton. It's not as abstract.

Three years later Worm, Lotze, and a roster of colleagues from all over the world published an even more controversial paper—Worm has taken to calling it the "crazy paper" because there's been so much fallout—which predicted a full collapse by 2048 of all fish currently commercially caught, unless practices change. This research paper has reached far past the narrow confines of the scientific literature, even spawning mentions on late-night talk shows, in newspaper cartoons, and in celebrity magazines.

Lotze and Worm are part of the young, fresh, and brilliant new generation of marine biologists who are trained to synthesize information, to draw conclusions. Katharina Fabricius and Joanie Kleypas are part of the same crowd.

Myers, who's older than Lotze and Worm, is a legend in marine science around the world for his passion, quirkiness, vast influence, and prolific work. Interviewing him is an exercise in shifting gears and speeding up. A few days before I am to arrive in Halifax, I get an e-mail from Lotze: Myers is in the hospital, diagnosed with malignant, inoperable brain cancer. It has happened catastrophically suddenly. The whole department, including the thirty graduate students Myers has been shepherding, is in mourning.

*

The story of humans and fish is the story of explosion in the human population and systematized extermination of fish.

Demographic analysis shows that 2,000 years ago, at the beginning of the Common Era, the human population on this planet was about 300 million. A thousand years later, it was nearly the same, perhaps 310 million. A thousand years after that it was six billion. Terrible plagues such as HIV/AIDS notwithstanding, the earth's human population is still growing at 75 million a year.

Lotze, who holds the Canada Research Chair in Marine Renewable Resources, can catalog how this has played out with fish over the centuries. It is a pattern of serial extinctions. When humans deplete one type of fish, we move to another, shifting over time from big, easy fish to smaller ones. We empty the streams, rivers, and lakes, then the shores, then the sea, and finally the open ocean, a trend that takes those who fish further and further from land and deeper under the surface of the sea. When humans can't easily find wild fish we figure out how to farm them, as we did with carp in the Middle Ages and as we're doing now with Atlantic salmon in many parts of the world.

When I meet Lotze in her lab, she starts telling me about her studies of the Wadden Sea. It's one of just twelve sites across the world so far that have had a full historical workup by Lotze and others using reconstructive detective techniques. Some of the others are Quoddy Bay on the eastern edge of Canada, Chesapeake Bay on the east coast of the United States, and San Francisco Bay on the U.S. Pacific coast. Of these, the Wadden Sea has the best historical records.

The Wadden Sea—formed 7,500 years ago when the Ice Age retreated and the sea rose—runs across part of the northwestern edge of Europe, flanking Denmark, Germany, and the Netherlands. It is the world's largest intertidal system, a shoreline that is exposed at low tide. In addition, it is one of the world's most important coastal wetlands, which is why it's a candidate for designation by UNESCO as a World Heritage Site, an honor reserved for the planet's truly magnificent natural sites.

These two characteristics—the intertidal system and the coastal wetland—are a recipe for biological variety. The Wadden Sea is one of those margins between ecosystems where the land meets the water and they both meet the air. Life here can exist on many levels of the food chain at once, each feeding another in a robust dance.

However, Lotze has discovered that, as rich as the Wadden Sea seems now, it was once almost unimaginably richer. Until about 1,000 years ago it was a fabulous, interconnected system that was home to gray and right whales, porpoises, dolphins, seals, astounding three-dimensional oyster beds, eagles, pelicans, flamingos, egrets, herons, and other large birds. Massive cod, haddock, halibut, and rays teemed in its waters, while meters-long salmon and sturgeon swam from river to sea.

Today the whales, including the grays that might once have had calving grounds in the Wadden Sea, have been extinct in this part of the ocean for centuries, hunted to the brink. Porpoises and dolphins are rarely seen. Gray and harbor seals are slowly coming back thanks to a reintroduction program in Britain. Many of the big birds are now uncommon or have vanished altogether, killed off for eggs, feathers, down, and flesh. Oysters with their huge, reeflike beds are almost extinct. Big fish have not been seen in rivers or the sea in many generations. Today, only three rivers in all of Europe still have sturgeon, once the favored European fish, which could grow to 400 kilograms.

As in so many other parts of the global ocean, whole, intricate communities of sea life are gone. Great swaths of the food web are no longer there, lost mainly at the top, and neither are key living spaces that once supported them, such as sea-grass beds and salt marshes.

What happened? The growing reach and sophistication of human society that began in the Middle Ages—including the burgeoning mercantile class, the growing population, and the early Christian church—had an unprecedented impact on the surrounding natural world. During that time people began to eat not just what was handy but also what was fashionable and what was allowed by the church. From about the fifth century, the Christian church forbade its faithful to eat "flesh"—that is, meat from land—on Fridays and holidays, about 130 days each year. In western Europe, including around the Wadden Sea, Christians were allowed to substitute fish—anything that swam in water, including what we know as marine mammals—for meat on those days. They didn't like it. Meat from the land was much more palatable to them.

Since the Neolithic period about 5,000 years ago, after the hunter-gatherer era ended, the human settlers of the Wadden Sea area had eschewed ocean food in favor of food they could find or grow on the land. If they had to eat fish, as the church decreed, they vastly preferred fish from freshwater to that from the ocean.

Richard Hoffmann, a historian at York University in Toronto who works with Lotze, has examined garbage heaps, menus, kitchen accounts and other archaeological and historical evidence showing that Europeans before the twelfth century ate pike, sturgeon, and salmon, all of which could be caught in streams nearby. By the twelfth century, though, these same records show that people were eating fish from the sea. Freshwater fish, or fish such as salmon and sturgeon that lived at least part of the year in freshwater, were still cherished above ocean fare, but they were becoming rarer and more expensive. Eating them became a mark of prestige, both in the world at large and within exclusive church circles. The rare pike, for example, became a favored fish of European prelates. For most Europeans, these prized fish were out of reach.

Once the local fish, whether fresh or marine, were fished out, a commercial fishery evolved to supply the need. At first it was small. In the Middle Ages this infant industry sometimes involved selling the extra fish from a nobleman's lakes or shores. Sometimes it meant drying, salting, or brining the fish so they could be transported to markets a little further away.

Fishmongering became a profession, driven by local extinctions of wild fish and by the requirements of the church. Hoffman's research shows that by the eleventh and twelfth centuries, fish had become a key commodity in the new exchange market system that was beginning in Europe. His examples from France and England parallel the development of the fish trade near the Wadden Sea. By the mid-thirteenth century, fast horses were taking marine fish from Normandy to Paris. Fifty years later, sea fish could be had even in central England, carried there by cart from English shores. By the fifteenth century the fish trade had expanded even further. The streams of Normandy had no salmon left, so fishmongers there imported them from Scotland and Ireland in barrels of brine and sent them to Parisian tables. Between the sixteenth and eighteenth centuries, cod, as well as other fish, vanished from the coastal seas around Europe, so intrepid fishermen went to Iceland to catch them, and eventually to the banks of Newfoundland in the New World.

Sturgeon had become even rarer. Archaeological digs from the seventh to the ninth centuries show that in parts of Europe sturgeon made up about 70 percent of the fish consumed by people. By the twelfth century sturgeon were nearly absent in streams from Italy to the Baltic. The few that could be found were small, no longer the 400-kilogram whoppers that once had swum the rivers of Europe. Cookbooks of the day offered recipes on how to make veal taste like sturgeon. Eventually sturgeon became so rare that they were designated the "royal fish," meaning that only royalty could consume them. Today, they're still endangered in most parts of Europe, and trade in sturgeon eggs—caviar, the modern luxury food—is tightly controlled.

Medieval Europeans not only created markets for wild fish when they could no longer be caught locally, they also made big business of farming fish; these were two key economic innovations of the age. By 1300 carp were being grown throughout Europe in large, man-made ponds. They were easy to rear, eating mainly garbage.

It wasn't just the fish and marine mammals that vanished from the Wadden Sea and the rivers that fed it. The living spaces of water creatures—and the birds that depended on them—were also systematically destroyed over time. It began during the Bronze Age, after 1800 BCE, when farmers cut down trees along the sea's edge for lumber, fuel, and fodder for their animals. By Roman times, near the beginning of the Common Era, the tidal forest of elm and ash was gone.

The Frisians, who settled on the Wadden Sea in the early Middle Ages, made an industry of salt. That involved digging up the tidal peat banks and the eelgrass beds, burning them, and taking salt from the ashes. By the end of the Middle Ages, the tidal bogs had disappeared. In the same era farmers began building strong dikes and drainage ditches to keep back the water for agriculture, essentially hand-making the landscape and getting rid of marshland. Residents of the area began harvesting sea grasses for mattresses and insulation, tearing up thousands of hectares of living structures of the ocean bed.

So the human inhabitants near the Wadden Sea fished or harvested much of the sea's life and eliminated the places those creatures could inhabit if they returned. The ecosystem that remains is far less complex, with fewer layers of life, and so is more vulnerable to other threats such as disease, higher temperatures, and lower pH. But because these changes happened so long ago in that part of the world, they have passed from the local human memory.

"In North America, people remember the big fish," Lotze tells me. "In Europe, it's much harder. Flounder are only this big." She holds her hands a few centimeters apart. "All the big fish are gone and somehow, nobody cares. People don't think they have an impact." Humans, she says, seem to think they are removed from these fundamental cycles of the planet, especially the ocean. In fact, the opposite is true. Everything we do every day is intimately connected with the ocean. As Worm puts it, every tear we shed ends up in the ocean, just as it originated there.

Lotze unrolls a huge, nine-panel map of northern Europe and the surrounding waters made in 1539 by the Swedish Catholic archbishop Olaus Magnus. He was a faithful naturalist and geographer, and this is the first detailed map ever produced of this area. It is a work of art, if not of unimpeachable science.

The map is also remarkable for the catalog it proves of what the ocean meant to humans of that time and place. It gives a glimpse into how our modern ideas of the sea were born. And how we got to where we are now, both practically and philosophically.

To Olaus Magnus, the sea was full of monsters. Whales were fierce, with claws, doglike teeth, green armor, double blowholes, and lizardlike snouts. In the illustrations that illuminate the map, a giant lobster clutches a hapless man in its pincer. A fearsome crimson sea snake crushes a ship. A tusked, boarlike creature with flippers and a fish tail boasts three blood-red eyes on its torso. Storms swirl in the air; gyres bigger than the mythical maw of Charybdis suck ships into the deeps.

The ocean was a place of danger and death, outside the scope of the civilizing force of humans. It was unpredictable, hellish, the unknown other; the enemy, a proxy for the unfathomable demons in the depths of the human soul. It was the opposite of the safe places that humans carved out of nature and subdued to their will and their god.

In *Description of the Northern Peoples*, the book Olaus Magnus published in 1555 as a companion to the map, he goes even further, devoting a whole chapter to the horrors of monsters of the sea, in which he includes fish such as halibut and flounder, a sign of how large the fish of his era were. It's an archive of what's gone. Today, Lotze says, the main commercial fishery the Wadden Sea is capable of supporting consists of shrimps and mussels. Even these are small.

*

Lotze and Worm are telling me about Ransom Myers. It's a painful subject. The cancer hit without warning in the part of his brain that controls his ability to express himself. The part that takes in information is still working as brilliantly as ever, and so possibly is the part that wrings meaning out of information. But he can't tell anyone about it. He can say yes and no and sometimes one other word, but no more. It's clear to me that if I want him to tell me the future, I will have to turn to the oracle of the papers he has already published.

Because it happened so abruptly, Myers didn't have time to write down all the new analysis he had done, all the synthesis brewing in his brain. It's the nightmare of a scientist whose life's work is devoted to warning of dangers to come. Myers is one of the few marine biologists in the world capable of understanding what's happening with fish, and it's all locked in his head, and a computer no one knows how to sort through.

It appears to be a metaphor for the fish he studies and so many of the other ecosystems of the sea: once something is lost, it can't be recovered. There is a point of no return.

*

Worm is a man under siege. He's just given a lecture to other scientists in his department on the "crazy paper," capping about three months of patient, unceasing defense of this work. It's the first major paper on the future of fish, and it predicts, among other things, the total collapse of all commercial fisheries by 2048 unless practices change. It essentially delivers a forty-year fair warning, dropped into a time when few realize there are any deadlines at all, much less one so soon.

The paper, "Impacts of Biodiversity Loss on Ocean Ecosystem Services," published on November 3, 2006 in *Science*, spawned more than a hundred newspaper editorials around the world, as well as a mention on Jay Leno's late-night television show.

Worm's audience today was friendly, because he works with people who are able to think in new ways about fish and who believe that this type of paper is a worthy first stab at a phenomenally complex subject. In other scientific circles the paper has been bitterly, emotionally contested—assailed by fisheries scientists and praised by marine ecologists. *Science* has published attacks from other fisheries researchers and rebuttals by the paper's authors.

Because Worm is the first author listed on the paper, he has taken most of the flak. He's unrepentant. He's been telling me that as a boy growing up in Germany he was crazy for maps, an explorer at heart. I think that's why he enjoys swimming against the scientific current if he believes the current is flowing the wrong way: it's part of his own intellectual exploration.

Worm is the unusual sort of scientist who believes in helping people challenge their assumptions. In his conservation biology class, for example, he makes a practice of asking a biological question to which there is no known answer, and then getting his students to research it and write papers on it.

On the wall across from his desk at the university he has posted several of his favorite newspaper cartoons, each based on one of his articles. In one, an old man says wistfully to a child, "In my day, we used to eat fish with our chips." Another depicts a museum exhibiting three extinct animals: a wooly mammoth, a dinosaur, and a fish. A third shows a fish on a psychiatrist's couch; the doctor tells it, "It's not your imagination. . . . The whole world *is* out to get you."

The impact of the paper's forty-year warning doesn't seem to wane. While we're talking Worm gets a phone call from a researcher at *Vanity Fair*. He's never even read it. They want him to help them put together some information on fish trends. He tells them yes and gets off the phone, shaking his head. Later I bring him a copy of the latest issue, and he laughs uproariously as he leafs through the stories of fashion, celebrity, and easy wealth. They have so little in common with his life.

In fact, the "crazy paper" did much more than posit a deadline for the salvation of fish. The figure 2048 was almost a footnote, tossed in to highlight the urgency of changing fishing practices. The larger message of the research is that the global ocean, as a whole functioning system, provides important services to humanity, protecting us from floods, removing toxins from waste, and serving up food. (In addition to other benefits, like regulating the planetary carbon and oxygen cycles, which were beyond the scope of the article.) As humans remove different types of life from the ocean's food web, we impair the ability of the global ocean to provide those services. And, therefore, to keep itself healthy.

For example, fishing out whole species and then whole families of sea life makes the life that remains far less likely to thrive. The more types of life there are in a given part of the global ocean, the healthier that part of the ocean is, and the more likely it is to be able to provide food and other services to humanity. The complex, planet-supporting ocean system that humans are just starting to understand has evolved to work in balance. When humans remove great swaths of species from the mix, things don't work the same way. The symbiosis of species and chemistry changes. The system becomes vulnerable. Fish whose populations have been fished out don't come back easily to their original numbers. Floodplains and salt marshes that have been destroyed can't filter the toxins from water, and that leads to more toxic algal blooms, fish kills, beach closures, and low oxygen levels in the water.

The "crazy paper" builds on the infamous "90 percent" paper, which in turn builds on Lotze's work about historical damage to the structure of coastal ecosystems. While Lotze's work looks at specific places where the ocean meets the land over time, the "90 percent" paper looks at commercial catches around the world since large-scale, intensive industrial fishing began in the 1950s, both at continental shelves and in the open ocean.

One of the main fishing mechanisms throughout this period was long-line fishing, which involves setting hooks at regular intervals across vast areas of the ocean. That means it's possible to compare over time how many fish have been caught for every 100 hooks. Figures from the open ocean come from Japanese fishing fleets from 1952 to 1999, because Japan had the most widespread long-line fishing operation, extending to all oceans except those at the poles. The findings were astonishingly consistent. During the first few years that a new type of fish is targeted, catch rates are very high. A few years in, the rate plummets.

In the Gulf of Thailand, for example, 60 percent of sharks, skates, and large fish with fins were taken from the ocean in the first five years of industrial fishing. In the narrow shelf of South Georgia in the southern Atlantic Ocean, fish stocks were depleted in the first two years. On average, 80 percent of the populations of fish vanished within fifteen years of the start of industrial activity. Overall, Worm and Myers calculate that the populations of every single large predatory fish across the global ocean fell by 90 percent in the fifty years after industrialized fishing began.

As Worm and Myers point out, that often occurs before any scientific monitoring takes place; by the time the scientists arrive to try to figure out how much fishing is sustainable, the damage has been done. The fish populations they are belatedly called on to maintain are already historically low levels. It doesn't leave much room for error.

As fishing became more difficult, other studies have shown that fishermen began spending both more time and more money to catch fewer fish, using sophisticated, expensive sonar and satellite equipment to target their catch. As well, they began to go yet deeper into the ocean and lower on the food chain. It's a recipe for trying to catch the very last fish.

One of Myers's PhD students, Peter Ward, has found that the commercial fishery now routinely goes to 400 to 500 meters, about double the depth it went to in 1953. Catches of deep-water creatures—mainly deep-ocean squids—jumped by more than 20 percent between 2002 and 2004, according to figures from the Food and Agriculture Organization of the United Nations (FAO). Moreover, a study published in 2006 in *Nature* showed that populations of fish that live in deep northwest Atlantic waters were decimated in just seventeen years once commercial outfits targeted them, with just 2 percent to 11 percent of the fish surviving.

Another study, by Daniel Pauly of the University of British Columbia and others, has calculated that the world's total fish catch peaked in the mid-1980s and has since declined. An estimated one-fifth of all the fish caught and killed in U.S. waters are thrown back into the sea because they're the wrong type. In other countries, that could be higher. Pauly and his research group have also shown that fishing fleets are using ever more fossil fuels as they travel further to catch fewer fish, one of the few major industries on the planet whose energy use is becoming less efficient. The commercial fishery now accounts for 1.2 percent of the world's annual oil consumption, using more than twelve times as much energy to catch fish as the fish provide in food energy to those who consume them.

How do fishermen pay for all this energy, equipment, and time? A study published in 2006 by Pauly and his colleague Rashid Sumaila shows that government subsidies to the global fishery ring in at between $30 billion and $34 billion a year, much of that going to offset the costs of fuel. As a point of comparison, Sumaila calculates that the total value of all fish caught in 2000 was about $80 billion.

The modern trends show the same picture that Lotze's research found at the thinned-out ecosystems where the ocean meets the land in the Middle Ages. First, fishermen go further out to sea to find the fish; then they switch to a species lower on the food chain; then they increase the technological level of their fishing gear while also expending more effort to catch fewer fish; finally, they turn to farming fish.

FAO figures show that in 2004 about a third of the fish humans ate came from farms, up from about 4 percent in 1970. That proportion is growing year by year as the wild fishery declines and becomes more difficult to fish. During roughly the same fifty-year period that Worm and Myers examined in their "90 percent" paper, aquaculture production in the world rose from less than 1 million tons to nearly 60 million tons. Aquaculture is growing far more rapidly across the world than all other types of animal farming.

In a cruel irony, one of the staples of aquaculture is small wild-caught fish, which are ground up and made into meal to feed the captive creatures. It takes 2.5 kilograms of ground-up herring to produce 1 kilogram of farmed salmon, Lotze has told me. On the bright side, that's down from the 4 kilograms of herring needed to produce the same kilogram of salmon in 1980.

What are the implications of all that Worm, Myers, and Lotze have found?

Worm takes the synoptic view. It took humans roughly 50,000 years to deplete the planet's large land animals, 5,000 years to exhaust most of the planet's coastal environments, 500 years to fish out the continental shelf, 50 years to impoverish the open ocean, and about 5 years to run through the creatures of the deep ocean.

"We don't realize how absolutely exceptional this time is. We are reaching the carrying capacity of the planet's natural resources. We are at the stage of losing the ability of things to come back on their own." He pauses, then says carefully, looking me straight in the eyes: "That would represent a threshold."

*

I've just come from a PhD thesis defense with Worm and all I can think about is apoptosis. The defense, by Ward, one of Myers's graduate students, was a roaring success. Ward can now officially call himself Dr. Peter Ward. He's jubilant. But as triumphant as his day was, it was also poignant. Myers, as Ward's supervisor and coauthor on several papers, ought to have been there, doing his usual shtick of appearing to pay no attention, banging on his computer keyboard, and then surfacing to ask the most piercing question of all.

In spirit, Myers was there: one of his other students was there videotaping the whole event to show to Myers in the hospital.

Ward has carried on Myers's work of gauging present fish populations in the global ocean compared to fifty years ago, when industrial fishing began. Five chapters of his thesis have been published as peer-reviewed journal articles and a sixth has been accepted for publication. Ward is already an established scientist working on the some of the most critical questions humanity is facing.

And he's got the battle scars to prove it. He routinely gets "battered and bruised" at international fishery meetings, he says, and it baffles him. Ward's research shows catastrophic declines across all commercial species of fish, with the fish that remain being far smaller than historic norms. He's trying to explain to the world community what this means—how fishery catch models need to change if modern catches are to be sustained. But time and time again he is told that his data have to be ignored because they don't fit the model.

Who makes the fish-catch model? Who knows. It's a black box, he says. He thinks it should be opened.

"We've never exploited these fish to this level and we don't know what happens next," he tells the defense committee.

So what is apoptosis? Cells in healthy bodies are programmed to kill themselves if they are damaged beyond repair, starving, poisoned, or diseased. It makes sense. The healthy body must continually make new cells—about 50 to 70 billion a day in adult humans—but to stay in balance, in homeostasis, a similar number of old cells must die. That's how the body stays within the narrow range of the biologically possible (and why you don't see humans with huge hands or one exceptionally large lung). It's also how the body stays healthy and recycles the useful components of its aging cells. It's similar to other inexorable cycles of biology: death leads to life leads inevitably to death. Winter must be followed by spring. Time is a loop, not a line.

So, when a cell gets to the point of no return, instead of trying to stay alive, it switches into apoptosis mode, dies, and can be replaced by another, healthy cell. Apoptosis, therefore, is both good and necessary. In many forms of cancer, however, sick cells lack the ability to self-destruct. They refuse to die, replicating instead along with the disease. That's the definition of a tumor.

This is another way to understand what's happening to the global ocean system. The system's chemistry is changing in ways not known for at least twenty million years; its life forms are being demolished at an unprecedented rate with unknown consequences; its fecundity is in question, and so is its metabolic rate. It faces continued assaults in the coming decades from yet more carbon dioxide, physical destruction, and overfishing unless human behavior changes dramatically. The vital signs are not good.

Is the ocean then edging toward apoptosis—not on the level of the single cell but at a systemic level—dying, casting aside the old system with its destructive humans so that it can create something new?

I go back to Worm. He is resolutely hopeful that humanity can change its destructive patterns. We can, he avers, learn from the past. He, Lotze, Myers, and many other scientists have put that belief at the core of their lives. It's why they study the things they do.

Part of the problem now, he says, is that what's happening in the ocean is hidden from the public. People don't know what's at stake. Take fishery catches, for example—why are we continuing to fish too much? Who's in charge? That's difficult to determine. The identity of crucial decision makers within the fisheries industry is obscured by the lack of corporate accountability. They are not public figures. They do not have to explain their decisions or defend them against public inquiry. They don't even tell the public what their decisions are. It's not much of a model. And it's plainly not working for the long haul.

Worm points to a better model, one that Myers used to praise lavishly. It dates back to December 2, 1946—the International Convention for the Regulation of Whaling. Established because all species of whales were in danger from, as the convention says, "overfishing of one area after another and of one species of whale after another to such a degree that it is essential to protect all species of whales from further overfishing," the convention has succeeded in preventing the extinction of all species of whale and in increasing their populations. Whales came to be seen as a global "great natural resource" that needed to be protected for future generations. So a committee now meets to decide whether any whales will be hunted and, if so, how many. The decisions are made public and debated. Data on how the decisions are working are also public, as are trends in whale populations.

Similarly, Worm says, the Convention on International Trade in Endangered Species of Wild Fauna and Flora, brought into international force in 1973, has been "100 percent successful." It regulates the trade of animals and plants that are known to be in biological danger so that their populations remain healthy. Nothing put on the list has gone extinct. Again, the governing committee's meetings are reported on and so are its decisions.

Worm points to a variety of other environmental solutions that have been implemented over the years, including controlling coal pollution in cities, DDT, PCBs, and dioxins. There has also been a successful international pact in 1987 to ban the use of ozone-depleting substances.

In Worm's home continent, Europe, there are now carpets of wind farms producing energy to replace fossil fuels. The land where Lotze grew up on the Wadden Sea has been turned into a solar energy farm by her sister's husband, a livelihood none of the earlier generations of farmers and fishermen who lived there over the millennia would have dreamed of.

"I'm absolutely hopeful," Worm says. "I would be suicidal if I weren't hopeful."

Lotze and Worm drive a hybrid car and have converted their home overlooking the Atlantic Ocean so that it draws heat from the sun and water from the roof. I've noticed that for lunch they eat homemade mackerel sandwiches—low on the fish food ladder—and that Lotze continually has half packets of sugar in her pockets, because she can bear neither to use a full packet in one of her cups of coffee nor to throw out the unused half. They are the smart new generation, both as scientists and as citizens. They have learned from the past and have retained hope. They have taught themselves to ask the questions that don't have easy answers.

Just as much as the fish they study, they are part of the life force of the future ocean. They have the ability to understand the ocean's secrets and tell us about them, to give us fair warning.

The month after I leave Halifax, I read in a newspaper that Myers has died.

Reading the vital signs: Medical history 7

SPAIN

The last time the sea was sick

The earth is a jealous guardian of her tales. She lays them down with exquisite precision and then hides them, building an archive only to lock it away. Finding the key is an uncertain business, dependent as much on chance as design. And when a text is found, reading the tales themselves—the secrets of the earth's past—takes yet more serendipity and skill.

A case in point is the PETM—Paleocene-Eocene Thermal Maximum, named after the geological epochs that bracket it—an elusive fossil marker from 55 million years ago, when the planet's atmosphere heated up dramatically. This slice of the fossil record holds evidence of a global crisis that affected the planet's systems, throughout land and sea, but only in the past decade have scientists begun to understand its lessons. To them it has become the Rosetta stone of global climate change, the best analogue available for what's happening in today's global ocean. In short, it allows them to ask a telling question: What happened the last time the planet was sick in this way?

So I find myself on a rainy morning climbing the Pyrenees, in the north of Spain, in search of the PETM. This is the field trip portion of a select international conference, "Climate and Biota of the Early Paleogene," held every three years or so for the paleontologists and climate scientists who are studying this deeply hidden piece of history. There have been four days' worth of cutting-edge scientific papers on the phenomenon, and now I want to put my hands on it, bear witness to what it means.

Henk Brinkhuis and Appy Sluijs, two paleoecologists at the University of Utrecht in the Netherlands, are the conference's current superstars. A few weeks ago they scored the cover article in *Nature* and hundreds of media hits around the world with their analysis of core samples from the deep Arctic, which showed for the first time just how warm the northern sea was. Many of the other papers we have been listening to in an auditorium in Bilbao are also destined to shake up our view of the world. This is a preview of science headlines to come and, as best as anyone can tell, of the future itself.

So what exactly happened during the PETM? It was already warm, unlike today, when the planet is in a cool phase. The concentration of carbon dioxide in the atmosphere was already high, estimated at anywhere from 600 to as much as 2,800 parts per million by volume. (Today, by contrast, the concentration is about 387 parts per million, up from about 280 before the mass burning of fossil fuels began a quarter of a millennium ago. Before that, the concentration had remained below 300 parts per million for at least twenty million years.) Then, in a geological instant—that is, less than a thousand years—an almost unimaginable amount of carbon was released into the atmosphere and ocean all at once. This carbon burp, similar in chemical composition to the residue from burned oil, gas, and coal, slightly more than doubled the atmospheric carbon dioxide concentrations. The carbon and oxygen isotopes contained in the fossils from this period tell the story.

First, though, a chemistry brushup. Carbon is one of 94 naturally occurring chemical elements on earth. (Another 23, so far, have been created artificially.) All atoms of carbon contain six protons, so it is given the atomic number 6 on the periodic table. This is invariable. But the nucleus of each atom also contains neutrons, which—in carbon as in most elements—vary in number. That means the atomic mass, or weight (the sum of an atom's protons and neutrons), also varies among atoms of the same element. These atoms of different weights are termed isotopes of the element. Carbon can have six, seven, or eight neutrons, which means that the atomic weights of its isotopes are 12, 13, and 14.

Plants and animals inevitably draw a mix of these isotopes from the atmosphere, but the lightest isotope, carbon-12, is the easiest for them to use. If there's a tremendous amount of carbon in the atmosphere, then, plants and animals will use more of the light carbon isotope, leaving behind the heavy. When fossils show more than the usual proportion of light carbon isotopes, rather than the classical balance of light and heavy, it means carbon was superabundant during the organism's lifetime.

Fossils from the PETM—whether from land or sea—show a stunningly sharp spike of the light isotope (in technical terms, a carbon isotopic excursion). The spike represents the addition to the atmosphere of as much carbon as would have been released if every living thing on the land or in the sea had been instantly incinerated. What is critical is not just the extraordinary amount of carbon but also the speed with which it entered the planetary cycle. It happened too fast for much of the living earth to adapt. The system that had existed was overwhelmed, pushed past its limits, and a new system was born.

From the tropics to the poles, the surface air temperature rose by at least 5 degrees Celsius in less than 10,000 years, a catastrophically fast increase. The ocean suffered its own chemical catastrophe. The carbonate compensation depth—the level below which calcium carbonate dissolves in the ocean—rose dramatically, a sign that the ocean's acidity also spiked following an influx of dissolved carbon dioxide from the atmosphere. Several of the ocean's currents stopped or shifted course, and the sea level rose. The coral reefs of the ancient Tethys Sea—which later, as the continents and tectonic plates shifted, became the Mediterranean and gave birth to the Atlantic Ocean—died off, replaced first by algae-slicked coral reef structures and then by brand-new large species of the primitive marine organisms called foraminifera.

The deep sea itself, which until recent generations scientists had believed to be immune from big changes, warmed by 4 or 5 degrees Celsius, and so did the surface waters of the hurricane-prone tropics. Dinoflagellates, a common type of marine plankton, spread north and south from their normal subtropical range, following the warmer waters. As Brinkhuis and Sluijs discovered, the Arctic became a giant warm bathtub with a freshwater lid. Surface waters there warmed by 9 or 10 degrees Celsius and were thick with fast-growing freshwater ferns.

Temperatures in the middle of the big continents of the era (which were different from those of today, as was the configuration of the global ocean) increased by an average of 5 degrees Celsius. In the tropical rainforests, annual average temperatures rose to as much as 34 degrees Celsius. (The threshold for plant death in those areas is 33 degrees Celsius.) One study showed that roughly three-quarters of the Colombian rainforest died. The planet was sweltering. It rained copiously. In the fossil record here in the Iberian peninsula there is evidence of extreme flooding, ferocious storms, and frequent cyclones, and those patterns may have been repeated elsewhere.

Across the planet, wherever paleontologists look, whether on the bottom of the ocean or on land, they see extinctions of creatures that had been hardy enough to survive the chaos that had killed the dinosaurs and so many other living things ten million years earlier. For example, as many as half of the species of foraminifera, one-celled creatures with chambered shells that dwell on the ocean bottom, suddenly became extinct, while other, larger foraminifera appeared for the first time.

Lots of other new creatures abruptly came into being, often miniature versions of species that had been plentiful before the great warming. It was, eventually, a splendid time for mammals, which proliferated and shifted into many new forms, moving across the continents to parts of the world that had been inhospitable to them before.

Paleontologists reckon it was roughly 150,000 years before the climate and ocean returned to the way they had been, making the PETM not only an intense and abrupt climate change but also a sustained one.

It's not just that the PETM, bearing the stigmata of intense global climate change, confirms the warnings that former U.S. vice president Al Gore, Australian biologist Tim Flannery, and much of the scientific community have been sounding. It goes further, revealing an aspect of the coming picture that is scarier than these warnings: Changes as vast as those now affecting the atmosphere and ocean are not linear. They feed on themselves and grow exponentially, reaching a point at which the system switches abruptly to one that is dramatically different.

Observers of the modern climate system are already seeing signs of unbelievably rapid change. Scientists charged with predicting how fast the permanent ice is melting in today's Arctic, for example, are having trouble keeping their presentations up to date because the melting is occurring so much faster than predicted. In the early 1990s there was talk of an ice-free Arctic in the summers by 2100. Seven years ago scientists reckoned it would happen by 2050. In 2007 a Russian ship made it through the Northwest Passage in September, when traditionally the ice is fast. Now, scientists are seriously examining whether the passage will be open all summer in just a few years.

At the same time, human emissions of greenhouse gases are increasing far ahead of calculations, as if they are feeding on themselves. A recent paper showed that while emissions rose at an average of 1.3 percent a year through the 1990s, between 2000 and 2006 they rose by 3.3 per cent a year on average. In part this is because the land and the ocean, which have been soaking up carbon dioxide, are getting saturated.

In the early 2000s analysts were predicting that China would become the world's largest emitter of carbon dioxide sometime between 2015 and 2020. In fact, that happened in 2006, when, for the first time, China outpaced the United States. This means that the global economy, of which

China is a major driver, is growing faster than expected, sending ever more emissions into the atmosphere and ocean.

Even if we stopped emitting greenhouses gases today, there is already so much carbon dioxide in the atmosphere that the average temperature would continue to rise by a tenth of a degree Celsius every decade for a century.

The latest report from the Intergovernmental Panel on Climate Change (IPCC)—a blue-ribbon report that all countries have to agree on before it can be published—has estimated that when the concentration of carbon dioxide in the atmosphere gets to 450 parts per million, which now seems both inevitable and imminent, the average surface air temperature will rise by about 2 degrees Celsius, and that will mean the extinction of from 20 percent to 40 percent of existing species on earth, a catastrophic pace. Many analysts say that the carbon dioxide concentration will not stop at 450 parts per million but will push on to 550, close to double the concentration before industrialization. At that point the average temperature will have risen by 3 degrees Celsius and as many as 70 percent of the planet's species will have gone extinct. That level of extinction has happened only five times before in the planet's history. The last time was when the dinosaurs died off 65 million years ago.

This is the portrait of a system poised to fall off a cliff.

*

The PETM looks unremarkable when I finally see it. We are at the Lizarraga Pass in the wild Pyrenees, part of the Navarre. The fog is heavy here, and it feels chilly after the wall of wet heat in Bilbao. Horses are strolling up a mountain path, nosing out favorite grasses; bells strung around their necks tinkle. Colts play. Over there a clutch of black sheep gaze down at us.

And here is the PETM. To me it looks exactly like everything else I'm seeing: just dirt. No obvious fossils, no rocks, nothing to distinguish it from anything else. Nothing I can make out to tell where it begins or ends. But to these geologists and paleontologists, this is a gold mine. They swarm it, pulling out hammers and magnifying lenses, and start hacking away, collecting pieces that they carefully label to take back to the laboratories of Europe and the United States.

Fifty-five million years ago, this was where the land met the sea, explains Robert Speijer, a geologist at the Catholic University in Leuven,

Belgium, who, along with Christopher Hollis, a New Zealand micropaleontologist, has agreed to explain the field-trip science to me as we go. (Speijer is the expert on the dismal fate of the coral reefs of the ancient Tethys Sea.) When the movement of tectonic plates forced the Iberian peninsula to crash up against Europe, all these geological records came to the surface from deep below, creating the Pyrenees. Thanks to that, and the sheep farmers who cleared the forests from this land in the Middle Ages, this part of the record is both visible and undisturbed, allowing scientists to read its history.

"This is the Paleocene," Speijer says, pointing to the fossil record on one side of the dirt in question. "And that is certainly Eocene." He points the other way, to a slice of earth that looks roughly the same. The chunk in the middle, then, has to be at the border between the two epochs, meaning it holds the stories of the warming that have so much to tell us about the fate of our own world. But Speijer and the others are not willing to assign this band, not quite 2 meters high, to the PETM without running a lot more tests on it. Thus the samples. They will take these clear plastic bags home, dry out the samples, extract microscopic marine fossils from them, and try to identify the minuscule fossils by species. And, hopefully, they will find the isotopic pattern that shows they lived and died during the PETM.

Speijer hands me his hammer and gestures to the seam of dirt where the other geologists are working. I climb over and bang off a piece of it. When I pick it up, it is soft and wet. It crumbles a little in my hands. I cradle it in my palms, thinking of the incantatory power of the archive I am holding, the mysteries it can reveal. When I show it to Speijer, he expertly breaks it into pieces, looking for signs of the fossils of ancient foraminifera—forams, as they're called in shorthand. The sample crumbles through his fingers and falls to the ground, now just dust to dust.

*

It has not been a secret that the climate might change abruptly. As far back as 2001 the Intergovernmental Panel on Climate Change sounded a warning. Scientists have tried, in the years since, to explain the threat to the public, but their findings have not captured the collective imagination. Perhaps it is hubris, that unlovely human characteristic, that has caused us to fail to perform due diligence on what the coming climate change will mean for humanity. We have, it seems to me, harbored the convenient fiction that somehow human ingenuity will make everything all right, that we are destined to survive, that we will pull a technological rabbit out of the hat.

Looking back on the IPCC report, I see that in its own jargony way it was strongly worded, if not popularly compelling. Scientists rarely deal in absolutes. They are trained to tease out the next questions to ask but not to publicize their conclusions in accessible and engaging ways. It was easy for the public and politicians to misinterpret the words as being less serious than scientists intended. This is from the summary document:

> Greenhouse gas forcing in the 21st century could set in motion large-scale, high-impact, non-linear, and potentially abrupt changes in physical and biological systems over the coming decades to millennia, with a wide range of associated likelihoods.
>
> Some of the projected abrupt/non-linear changes in physical systems and in the natural sources and sinks of greenhouse gases could be irreversible, but there is an incomplete understanding of some of the underlying processes.

That was alarming enough for some scientists to press for further investigation. In the United States the National Research Council set up a committee on abrupt climate change, which came out with a report in 2002 entitled *Abrupt Climate Change: Inevitable Surprises*. The report seizes on a single idea, the "surprising new findings that abrupt climate change can occur when gradual causes push the earth system across a threshold." It's the proverbial straw that breaks the camel's back. At that point, the climate sets off on its own path, at its own pace. The report concludes:

> Large, abrupt climate changes have repeatedly affected much or all of the earth, locally reaching as much as 10 [degrees Celsius] in ten years. Available evidence suggests that abrupt climate changes are not only possible but likely in the future, potentially with large impacts on ecosystems and societies.

By 2003 that scenario had prompted the U.S. Department of Defense to commission a study, which was completed that October. The report, *An Abrupt Climate Change Scenario and Its Implications for United States National Security* by Peter Schwartz and Doug Randall, focuses on what would happen if the Gulf Stream, which transports heat from the equator to the North Atlantic Ocean, shut off. Its predictions include shortages of food, water, and energy that "could potentially de-stabilize the geo-political environment, leading to skirmishes, battles, and even war. . . . Unlikely alliances could be formed as defense priorities shift and the goal is resources for survival rather than religion, ideology, or national honor."

The Pentagon report is, in essence, a plea for permission to plan for anarchy. Could this really happen? it asks. The answer follows:

> Ocean, land, and atmosphere scientists at some of the world's most prestigious organizations have uncovered new evidence over the past decade suggesting that the plausibility of severe and rapid climate change is higher than most of the scientific community and perhaps all of the political community is prepared for.
>
> If it occurs, this phenomenon will disrupt current gradual global warming trends, adding to climate complexity and lack of predictability. And paleoclimatic evidence suggests that such an abrupt climate change could begin in the near future.

*

A beautiful irony is built into these reports. The specific piece of the climate picture that spawned them is the rapid melting of the Arctic ice, caused by global climate warming. That phenomenon is pouring freshwater into the top of the Atlantic Ocean, diluting its salt levels. The less salty water that results is lighter than heavily salted water and doesn't sink as easily deep into the ocean. And that reluctance to sink disrupts the vast conveyor belt, known as the Gulf Stream or the Greenland pump, that carries heat from the equator to points north.

This conveyor belt is thought by some to be so critical to the global climate system that they have dubbed it the planet's Achilles' heel. Ice core samples from Greenland show that the Gulf Stream has indeed shut off, once 12,700 years ago during an event called the Younger Dryas and again 8,200 years ago. The theory is that if it shuts off again, the ocean-borne heat that keeps northern Europe warm will stay south, plunging England and parts of Europe into a cold spell. This is the scenario taken to its extreme in the 2004 Hollywood movie *The Day after Tomorrow*. It depicted a snap ice age in North America, complete with an ice-bound Statue of Liberty and valiant Americans freezing to death in their steps.

While the Gulf Stream has shut down before and is likely on its way to doing so again—there is plenty of evidence that the salinity is dropping sharply in the North Atlantic and rising in other parts of the global ocean—many scientists now believe this has been a red herring, drawing attention and research from far graver and more immediate threats. In fact, it's been called the equivalent of an urban legend. Matt Huber of Purdue University in Indiana, one of the world's top climate modelers, is helping me sort through the significance of the scientific papers presented at Bilbao. He says that most climate models show that winds drive the heat circulation, not the Gulf Stream. That's why, for example, Juneau, Alaska, is relatively balmy despite the lack of a similar ocean conveyor belt in the Pacific Ocean. "When you measure how much heat is transported [through the Gulf Stream] past 60 degrees north, it's indistinguishable from zero," he tells me.

The irony is that the Gulf Stream issue, which caught the attention of at least some policymakers and put the concept of abrupt climate and ocean change at least on the fringes of the map, was raised before the arrival of even more worrisome evidence about the phenomenon as a whole. That's the evidence just now emerging from these microfossils of the carbon-glutted PETM. Now scientists are asking, Where did the PETM's massive infusion of carbon come from?

One of the leading theories is that the carbon exploded in a gaseous burp when methane hydrates stored on the ocean floor somehow broke free. I'm reminded of all that methane sitting now at the bottom of the Arabian Sea and the concerns of Peter Burkill at the Plymouth Marine Laboratory that it will breach the surface.

Huber explains that methane hydrate molecules, when cold, are like geodesic domes, stable structures that remain in the ocean's deeps. But the bonds within these structures are fragile. If the molecules are shaken up or warmed, the bonds break and bubbles of flammable methane (a carbon compound and a highly efficient greenhouse gas) shoot through the water column at phenomenal speed. If that happened on a large enough scale across the ocean floor, intense global warming would happen instantly.

But what could have triggered such a methane burp? It could be that the deep ocean warmed by a couple of degrees. Or a landslide might have ripped up the slopes of an ocean basin and shaken up the methane hydrates. Or the carbon could have come from somewhere else.

Mark Pagani, of Yale University's Department of Geology and Geophysics, is an expert on historic carbon concentrations in the atmosphere, and he doesn't buy the methane theory. "It was a very sexy idea, and it's had its day," he tells me.

One problem is that it is not known whether enough methane hydrates existed in the oceans of 55 million years ago to spark the kind of warming that the fossil record clearly shows. Another issue is that, while the type of carbon contained in methane would warm up the world far more quickly than carbon dioxide, it would persist in the atmosphere for only about a dozen years before being transformed into carbon dioxide. The PETM lasted around 150,000 years.

Today, the isotopic signature of most of the extra carbon in the atmosphere shows that it comes from the burning of fossil fuels. In addition, as humans fell forests and till the soil, some of the landscapes that once absorbed carbon can no longer do that as effectively. "We don't know what the final result of this is going to be," says Pagani. "We don't know the consequences of our actions. The thing about the climate system is, it's a monster. Once it shifts, it's not going to shift back."

He pauses to think for a moment. Then, as if to scotch any sense I might have that he's an alarmist, he adds, "I'm not willing to say it's going to be bad. It's going to be different." He pauses one more beat: "Florida will be gone."

*

So where will it end up? And when? And how will *Homo sapiens* fare?

Huber, as the ranking climate modeler at this conference, is the professional seer of the future. While some of the scientists here are talking about catastrophe within fifteen years, Huber is comfortable talking about the next forty-five.

Sure, he's worried about the methane hydrates, the possibility of a permanent El Niño weather pattern that will turn the Amazon rainforest into a desert and give us an open Northwest Passage. He's heard other scientists talk about New York as New Atlantis, about Bangladesh submerged. Privately he figures that the human population is going to take a hit— "I don't see us getting away with any less than a very large population downturn"—but he'll only say that it will happen "in the next thousand years."

What's really disturbing him right now, though, is his theory about cyclones. The way he sees it, tropical cyclones are one of the mechanisms nature has developed to mix the hot waters on the surface of the ocean with cooler waters underneath. As they mix they move, ferrying the heat to new parts of the planet. Mean surface temperatures in tropical oceans have risen about a quarter of a degree Celsius in recent years. At the same time the intensity of cyclones has doubled. Huber's models predict that tropical sea surfaces will warm up by 2.5 degrees Celsius over the next century. The implications for the intensity of hurricanes are enough to make his upper lip sweat. Hurricanes of this magnitude are not now represented in any climate models, past, present, or future. But he can see how they could explain some of the strange phenomena of the PETM, including the fact that parts of the fossil record look like violent dumps of sediment.

Peter Wilf, a geologist at Penn State University, has other concerns as well. In the period preceding the PETM, the globe was already warm. The creatures then alive had evolved in warm conditions. Today, as carbon dioxide gets pumped so furiously into the atmosphere, we are in a relatively cool period of the earth's history. And the creatures alive today—including humans—are used to this cool regime. The PETM was marked by extinctions plus (over the very long term) creations. So what will happen during the coming abrupt changes? Will today's living things be as able to adapt to warmer conditions?

At the very least, there is likely to be more migration of plants and animals than there has been already. But much of the living space that animals and plants could once have moved to has now been transformed for human use. That could mean a lot of abrupt dead ends on the migratory paths. This time, the pulse of extinction might happen without the pulse of new life.

*

We are even deeper into the Pyrenees now, on our last day of tramping the mountains. We've been climbing for more than an hour along the treacherous banks of Les Compes gully, a tributary of the Esplugafreda ravine. Wild lavender, oregano, and thyme scent the torrid air, crushed by our steps. It's 38 degrees Celsius, midday and cloudless. Some of us are near heat exhaustion.

Fifty-five million years ago this was a coastal plain, probably partly arid. The internal evidence of the site suggests it had long, dry summers and huge cyclones and storms, formed while carbon concentrations in the atmosphere were rising. Sediments were laid down far more quickly than at other parts of the PETM, an effect of the energy of frequent storms and floods, says Birger Schmitz, a geologist at the University of Lund, Sweden, who has been working at this site for years.

Huber's cyclone theory takes on a new significance for me.

Again the paleontologists set to, hacking off samples to take home. They wrap the soil in wads of toilet paper and sheets of the *El Mundo* newspaper, like mummies, to dry them out and keep them safe until they can be analyzed back at their laboratories.

Off in the distance, climbing steadily in his hot plaid shirt, is a man I had assumed all along was another scientist. Last night I discovered he is Ron Waszczak, a senior staff stratigrapher with the Houston-based oil and gas firm ConocoPhillips. His job, he has told me, is to find new oil and gas reserves for his employer. He's here because of the tantalizing stories Brinkhuis and Sluijs published in *Nature* about their Arctic expedition and the fact that so many fast-growing plants lived there 55 million years ago. To him, they weren't stories about the planet's potential for catastrophic change. They were stories about where to look for source rock for the next fossil fuel reserves.

Reading the vital sign that is China 8

HAIKOU CITY

Slouching toward apoptosis?

As goes the global ocean, so goes life on the planet. This is beyond dispute. But here's an intriguing possibility. What if a corollary equation reads this way: As goes *China*, so goes the global ocean?

If that's true—and it's shaping up to be—then life on the planet will hang on the economic and energy policies China chooses and helps promote to other fast-growing nations such as Brazil, India, and Russia over the next two decades.

Despite its enormous importance, though, what China will do is a mystery. The People's Republic has policy plans for economic growth, population growth, fishing, aquaculture, and control of carbon dioxide emissions that point in two ways at once: toward a healthy future for the world and toward a very sick one. Either is possible. No one—not even the Chinese, and certainly not Westerners—knows which path the country will follow. China is the planet's great carbon enigma. It's important, when reading the sea, not to underestimate the power of this ancient and sophisticated people.

The idea that the planet's future might come down to China's choices is hard for people in the West to fathom. Westerners are used to thinking of China as a backwater, a view that the lavish summer Olympics in Beijing in 2008 was aimed at undermining. And China has had little to do with the centuries of industrial activity that have pushed up levels of carbon dioxide in the atmosphere and changed the ocean's chemistry. China's cumulative effect on the buildup of carbon dioxide has been minimal compared to, say, Europe or the United States.

However, China has vaulted upon the world economic stage with ferocity, and at a time when every molecule of carbon dioxide matters. This is the moment of truth in the international drama of fossil fuels and carbon emissions, the point where the actors decide how humanity will cope with carbon pollution in the atmosphere and ocean. The Chinese economy is growing faster than that of any other country—by a huge margin. Its population, already the largest in the world at 1.3 billion, is leaping upward by about thirteen million a year. It emits more greenhouse gases into the atmosphere than any other country and is also by far the planet's biggest player in the fishery and aquaculture, both of which affect the ocean's ability to keep itself healthy.

Consider the numbers: China has become the biggest emitter of greenhouse gases in the world, surpassing the United States in 2006, according to the Netherlands Environmental Assessment Agency. This has happened far more quickly than the international community expected. And China's emissions are continuing to grow quickly, mostly from burning coal, the fuel that gives off more greenhouse gas than any other.

This is the result of China's extraordinary, sustained economic growth. From 1991 to 2005 China's rate of growth averaged 10.2 percent a year. The report *One Lifeboat: China and the World's Environment and Development*, by Arthur Hanson and Claude Martin of the International Institute for Sustainable Development (IISD), says that China's economy is expected to continue to grow by at least 8 percent a year into the future. That would give China the third largest economy in the world by about 2015. Even Japan's economy during its phase of phenomenal growth increased at an average yearly rate of just 3.85 percent from 1971 to 1991. In 2007 China's gross domestic product grew by a stunning 11.7 percent, according to a report by Worldwatch Institute. That accounts for a third of that year's increase in gross world product. In all, notwithstanding recent embarrassing recalls of several toxic consumer goods, the Chinese gross domestic product grew nearly six times as much as that of the United States in 2007.

This astonishing growth has come from China's successful bid to become the world's factory. Citing the newsmagazine the *Economist*, the IISD report notes that China makes nearly a third of the world's television sets, half of all cameras, seven in ten photocopiers, and almost a third of the furniture. All that production, of course, requires vast amounts of many different raw materials. China, the report says, is now "the second largest consumer of primary energy after the United States, and the top global producer of coal, steel, cement and 10 different kinds of non-ferrous metals. . . . In recent years, due to its robust growth, China has replaced the United States as the dominant market and price setter for copper, iron ore, aluminium, platinum . . ."

This trend has become so pronounced so rapidly—and so unexpectedly—that China's economic might and multibillion-dollar trade surplus with the United States have become political fodder in America. And almost every bit of this torrid economic growth depends on the use of fossil fuels—mainly coal, the dirtiest of them all—and therefore on climate- and ocean-altering greenhouse gases.

Against that backdrop it's easy to think of China as a nation whose policies support economic growth at all costs. But will China, coming of economic age now when the perils of carbon are so apparent, do what other developed countries have done and heedlessly push the world closer to catastrophe?

To get a perspective on a piece of China's strategy, I sign on for an immersion course on the state of the Asian seas as a delegate at a high-level international marine meeting in Haikou City, Hainan Province, an island in the South China Sea.

It's a groundbreaking meeting of sorts. The Second East Asian Seas Congress is aimed at encouraging countries that border (or nearly border) the East Asian Seas—including China, Vietnam, Thailand, the Philippines, Japan, the Koreas, Indonesia, East Timor, Laos, and Cambodia—to take a sharp look at how they are managing the health of the coasts and ocean around them, and to try to coordinate policies. This is where science meets the lawmakers.

The congress's guiding premise is that the East Asian Seas, one of the richest, most diverse, and most degraded of the world's ocean ecosystems, is a communal resource in a lot of trouble, and that the roughly two billion people who live near it—fully a third of the planet's human population—must put aside historic differences in order to preserve its ability to provide food for them all. It's a simple enough concept, but international cooperation over marine bounty has rarely been achieved. A classic example of failure is the European Union's current intractable mishmash of fishery policy.

In the East Asian Seas, though, things are reaching a critical level. Nearly 90 percent of the region's coral reefs are threatened and half are already in dreadful shape. Seventy percent of the region's mangroves—a critical nursery for both land and sea creatures—have vanished in the past several decades, mainly cut down for short-term shrimp farms whose products are shipped to the West for cut-price sale. It's estimated that at this rate the mangroves throughout East Asia will have vanished by 2030. That represents a third of the world's mangroves. The loss of both coral reefs and mangroves means that habitat for the most vulnerable marine life—including the hatchlings and embryos that grow to become fish—is also gone.

Rivers and coastal waters in this region are heavily polluted, mainly because so many humans live close to the water and use it as a dump for chemicals, excess fertilizers, and other substances they don't want. Some rivers have become lifeless; dead zones, bereft of oxygen, are forming off the coasts.

The Asian fishery is in trouble, too, just as it is everywhere else. A speaker from the World Bank explained that it takes more than three times the effort to catch a fish in the Philippines today that it did at the beginning of the 1990s, a pattern repeated throughout the region. Chinese waters are so fished out that the government put a moratorium on summer fishing throughout its waters in 1998 and is trying to buy back fishing vessels. Other studies show that the fish still caught in the East Asian Seas are farther and farther down the food chain. But the numbers on how many fish China has taken over the years are suspect, so the actual catch trends there are not clear. In fact, inflated statistics about China's fish catch may have deluded the international community for years into thinking the global fishery was healthier than it actually was.

As for aquaculture, it is both the gold mine and the tragedy of Asia. As the global fishery has declined, and as demand for fish has grown, aquaculture has taken off. In 1970 about 4 percent of the fish and seafood humans ate came from farms; by 2004 it was a third. Aquaculture is growing faster than all other types of animal farming across the world. More than 80 percent of that farmed seafood comes from Asia, and more than half comes from China alone. China has said it wants that percentage to grow.

However, farmed water animals need healthy places to live. So far, apart from a few vaunted experiments, aquaculture in this part of the world has created profoundly unhealthy living places, degrading water and land so that it cannot be used for other plants and animals. For example, not only have the life-giving (and cyclone- and tsunami-breaking) mangroves vanished from the coasts in favor of shrimp farms, but those farms have also pumped untreated shrimp waste back into the coastal waters.

As Professor Mohamed Shariff, an aquaculture expert from Putra University in Malaysia, told the conference, sometimes the shrimp pollution—which he described as espresso-colored sludge—can spread as much as 3 kilometers offshore. Filled with phosphorus and nitrogen, it feeds toxic species of phytoplankton, which gather in blobs and poison shrimp and fish. To keep their harvest alive, farmers then invest in antibiotics if they can afford them; these remain in the flesh of the animals and in their surroundings. Eventually the shrimp ponds can no longer produce. It's the marine equivalent of slash-and-burn agriculture. Not only that, but the vast majority of farmed shrimp in China are *Penaeus vannamei*, imported in 1996 from Latin America without much research. These shrimp need less protein to grow fat than other shrimp and can live in both salt and fresh water, but they are vulnerable to Asian viruses as well as those from their home waters.

It strikes me, sitting in the opulent Haikou City People's Conference Centre, that this type of aquaculture seems like a case study of an enterprise fated to fail. Who thought this would work?

Yet it's clear to many scientists and fisheries experts around the world that fish farming is the only way to ensure that the world's population will have enough animal protein—the wild fishery will not withstand the level of catch that has been going on for the past decades. The trick will be to make sure the farming is done in an environmentally healthy, sustainable way.

I've seen far better models for aquaculture that could be implemented here, the world's fish-farming epicenter. I think of Phil Hart, a former police sergeant, whom I met in Whyalla, South Australia. He was one of the founders of a company that in 1998 became the first in the world to farm kingfish, a firm, white-fleshed native of Australia that is superb for sashimi. He took me around to the fish cages in Fitzgerald Bay in the northern part of the Spencer Gulf to show me the operation. These tide-drawn emerald waters, which face south to the South Australia Basin and into the Indian Ocean, are not pristine. But there's no agriculture and few humans live nearby. There's no chemical or organic waste being poured into the fish's living space.

Hart groaned slightly as he told me about the hoops he and his partners had to jump through to set up the operation, but he's proud of having met the tough standards and of how tightly the state of South Australia regulated the industry. It's the best example I've seen of science informing policy on a marine issue. The fish the company grows have to live naturally in that particular part of Fitzgerald Bay. It can't bring in foreign brood stock the way shrimp and salmon farmers all over the world have done; experience has shown that fish tend to get sick if they aren't in their natural habitat. The Asian practice of clearing the natural habitat is forbidden; cages have to be far enough apart and have few enough inhabitants that the sea-grass beds underneath remain undisturbed. As a result of the healthy environment, Hart's enterprise had only had to use antibiotics on the fish twice in nearly a decade of farming.

Reporting to the government is intensive and ongoing. The company—which had just been sold to Clean Sea Group, one of the largest industrial fishery and aquaculture groups in Australia—has to take samples of plants and animals on the seabed underneath and surrounding the cages. It is required to have plans to prevent farmed animals from escaping into the open sea, and to make sure dolphins, seabirds, and sharks aren't getting caught in the nets. It also has to prove they aren't feeding the fish too much and fouling the bay. Hart told me that environmental inspectors come around with clipboards to verify results and hold the fish farmers to what they say they're doing.

Ian Nightingale is the government official who administers the state's aquaculture legislation. When I left Whyalla and went to Adelaide to see him, he told me that his division did a great deal of scientific research on Fitzgerald Bay before it would allow Hart's farm to be built. Scientists produced conservative figures for how many farmed fish the bay could carry, and Nightingale has not let them be exceeded. The farm also could not interfere with the movements of endangered sea lions in the bay. In addition, the aquaculture legislation provides for both a license and a lease for any fish farm; it gives the government two ways to shut down an operation that isn't up to snuff. The state's independent environmental protection agency can also pull the plug if it doesn't like the results of environmental reports from the farms.

Nightingale knows that some fish farmers approach the business as a gold rush, but that's not what South Australia decided to do. He said that the state wanted fish farming to be a viable industry for the long term to replace some of the small land-farming operations that had vanished. The day I was there, Nightingale was preparing for a visit from fish-farming officials from Tanzania. Officials from China had already come by to look, in addition to those from Japan, Indonesia, and Vietnam.

*

Back in Haikou, the behavioral calcification that thwarts the efficacy of information sharing hits like a brick. Clearly a great deal of thought has gone into this conference. It's thick with policy makers, politicians, nongovernmental organizations, and scientists. The whole of Haikou City feels as though it's on high alert in recognition of the honor. I wake up at my hotel every morning to the rhythmic sound of women in pointed bamboo hats and flat black cloth shoes sweeping the streets that lead to the conference center.

Yet, the litany of marine concerns discussed at the conference—dropping fish catches, local water pollution, off-shore dead zones, destructive aquaculture practices, contested claims to the riches of the sea—are the traditional ones. They are important and even alarming, but they remain rooted in an ancient human understanding of the ocean, seen largely from the shore or from the surface. The ocean's depths—99 percent of the living space on the planet—their role in the world's life-giving chemistry, and how humans are changing that are not on the agenda. The world's policy makers have not yet grasped how important these issues are. Global climate change comes up from time to time, but mainly as a separate problem, one that will raise the sea level, take away precious human living space and livelihoods, and steal financial assets from those close to the shore.

Through 39 workshops and seminars, 285 papers and posters, a ministerial forum, and a youth event, the idea that carbon dioxide emissions are chemically affecting the ocean comes up just once, and that is in a half-hour session by Michael Depledge and Michael Kendall, both based at Plymouth Marine Laboratory in England. They are fluent, even passionate, in describing the fallout from the coming acidification of the ocean. Depledge argues for "climate change" to be renamed "ocean change" and warns of potential bursts of carbon that could create an alternate and less hospitable global system: the changing Indian monsoon that could release a burp of methane, the melting permafrost that could do the same thing.

Kendall gives an elegant talk on the consequences of acidic seas, raising the specter of a shift in the whole marine food web. "All these changes have happened before, but never, ever as rapidly," he says. "These are the fastest broad-scale changes the marine system has ever experienced." He is almost pleading at the end. "We can't just sit on our hands and wait," he tells the small break-out seminar. Climate changes are already running ahead of projections. In the United Kingdom the scenario predicted for 2020 came to pass in 2006.

It seems plain to me that all the problems people are talking about here in Haikou City and trying to find solutions for will be compounded by the chemical changes now happening in the ocean.

*

The day after the conference, as officials from the East Asia Sea countries are hammering out a formal statement agreeing to meet again in another three years with some more ideas on all these issues, I get yet another reminder about the entrenched nature of humanity's ancient, parasitic relationship with the sea. With Mark Wunsch, a marine biologist from the Phuket Aquarium in Thailand, I hire a car to drive 300 kilometers down a perfect, wide highway to Sanya, a resort town on the southern tip of Hainan Island, jutting into the South China Sea.

Sanya is home to a rather famous beach, a big tourist draw for China's growing middle class. Much of the tourist fare is fish-related, or made from shells, including inflated and dried blowfish sporting painted faces and small straw hats. We see food vendors offering to grill shrimp. The shrimp are threaded, live and still wriggling, on bamboo skewers. Wunsch points up the road to a shrimp farm he noticed on the way in. It's probably where these animals came from. I can only think of espresso sludge.

We are headed for Sanya's aquarium. Right inside its entrance is a shallow pool with an assortment of sea turtles swimming around. At the sight of us, one of the aquarium staff jumps to attention. Then he kneels at the side of the pool, drags one of the turtles out of the water, sits on it to keep it in place, and stretches its neck out as far as it will go. He gestures to a plastic-covered photo of a smiling Chinese woman perched on top of this turtle or another, indicating that we too could have our picture taken for a price.

He tilts the turtle's mouth up, posing it for us. We walk away.

All species of sea turtles are endangered, many critically, despite the efforts of the world community to protect them from trade and aberrant exploitation like this. Their ancestors have been on the planet for more than 100 million years, surviving even the extinction of the dinosaurs. I think of the endangered green turtle I watched swimming on the vast expanse of the Great Barrier Reef, its movements unhurried, poetic.

Inside the aquarium building are the fish tanks. In a few, sprigs of plastic plants stand out against the backdrop of swimming-pool tiles. Wunsch is particularly interested in comparing these displays to those at the Phuket Aquarium. The fish are hard to see, labeled poorly if at all. It's not clear which ecosystems they come from; the groupings seem random. While some look healthy, others clearly have skin problems, which is a sign of disease. A fish in one of the tanks is swimming on its side near the surface—not a good sign. In another tank, all of the dozens of inhabitants are gathered at the surface, gasping as if for air. Wunsch says that there is probably something wrong with the water's chemistry. He figures they will all be dead by morning. You can't mess with those biological limits.

A woman who works at the aquarium ushers us bossily out of the building and on to the next big attraction: the crocodiles. There are at least thirty sleepy-looking crocs in and around a concrete pond filled with soupy green water. The smell is enough to make me gag. The pond is surrounded by an elaborate platform, complete with what looks like a disk jockey's setup.

The turtle wrangler appears again, this time with a baby croc whose jaw is taped shut so we can pose with it for pictures. I look away, wondering why a stringy brown chicken is on the platform, pecking away at the remains of a watermelon rind in a Styrofoam container. Then the wrangler abandons the baby croc, swoops down, and picks up the hen, which squawks vociferously. He makes as if to toss it to the crocodiles with one hand, rubbing his other thumb and two fingers together in the universal symbol for money. We refuse the live sacrifice and escape further along the croc pond, away from the platform.

Moments later a group of dozens of exuberant tourists comes through. Almost at once music starts up at the console, piped through impressive speakers. A guard comes over and shoos us away. If we don't pay, we don't get to take part in the fun. As we walk away, a drum rolls. I look back to see the crocs, which had been sunning themselves on the concrete, slide deftly into the water. They know what's coming.

On the long drive back to Haikou City, I have plenty of time for reflection. This was not an experience specific to China. Other countries have dreadful aquaria as well. What I saw in Sanya is not that different from the awful little zoo I visited in Kuranda, Queensland, in northern Australia. There, they have a staff of koala bears that tourists can cuddle and pose with, for a price. True, as the human staff make a point of explaining, the koalas can only pose for 30 minutes a day, and only for three days in a row before having a day off. It's a better deal than Sanya's sea turtles and crocodiles get, but it amounts to the same thing.

In microcosm, this is the way humans have traditionally related to animals, to nature. We have seen them in terms of what they can do for us, without regard to the damage we do to them. We haven't understood the whole picture, just little, select slices of it. We haven't known the dangers we run. And once we become aware of what damage we're doing to the whole system? Will that affect government policy, not to mention individual action? Or will humans just keep doing what we're doing and watch the system slouch toward apoptosis?

China does provide some clues. While it is clear that China is far from a paragon on all environmental fronts—including marine—and that it may not yet have figured out the future effects of ocean chemistry change, some of its recent policies have posited it as an international leader in green development. In a sense, because it is starting from a relatively low level of development, China has a free hand to be truly innovative.

Unbound by guilt over global climate change because it played so little historic role in it, the Chinese central government appears to see clearly what the science is saying. It also grasps that the key to the continued growth it seeks is greater efficiency in the use of expensive natural resources. Waste is a barrier to growth. This is about survival in a rapidly changing world.

China can already track some effects of climate change on its own lands and coasts. By the central government's calculations, the average temperature has risen more than the global average; starting in 1986, for example, China had twenty consecutive warmer-than-average winters. Climate change is also causing droughts and floods in China; glaciers are in retreat all over the country; water quality is appalling; and the sea level is rising slightly more on average than in other parts of the world.

The result is that the Chinese central government has come out with some muscular policies on both energy and climate change, putting the policies of many developed countries to shame. Embedded in the publicity around them is this fascinating paragraph, likely approved by the central government, from a December 2007 article in *China Daily*: "In China, environmental protection and the conservation of energy and other resources are of paramount importance. It indicates how civilized a country really is." Given the disbelieving public statements on climate change, even in recent years, from political leaders in the United States, Australia, Canada, and other countries, this is an astonishing position for a major world power to take. It is almost a rebuke.

So it seems China is staking its national pride on its green policies. And perhaps more: it appears to be keenly aware of its international responsibilities, of the fact that the fate of many other creatures depends on what it chooses to do. It's impossible to know for sure how committed China is to these policies. The leaders could be posturing, and certainly these are difficult policies to enforce. But China has already made astonishing progress in reducing the amount of energy needed to expand its economy. While its economy has grown in recent years by about 10 percent a year, its annual energy use has grown by the comparatively small 5.6 percent. In international terms, this is a feat. China has also gone far further than the North American recycling system, adapting the policies of Germany and Japan to form a "circular economy," in which one industry's waste becomes another's raw material.

As well, China considers its controversial population policy part of a successful green plan. The one-child policy has held China's population to about 1.3 billion, rather than the 1.5 billion it would have been without the policy. That means the Chinese are using fewer of the earth's resources today than they might have been—the ultimate green policy.

China says it plans to go much further down the green road. Its list of climate-change policies reads like the manifesto of an environmental nongovernmental organization from the West. They include integrating climate-change policies with national social and economic policies—an intention frequently affirmed by Western powers even as they fail to implement it—focusing on both preventing climate change and adapting to what's coming, using science and technology to find new ways of cutting down emissions, and investing heavily in renewable energies such as wind and solar. Among the country's short-term goals is to reduce energy use by a fifth for every unit of gross domestic product by 2010, a bold move and one that other countries could learn from.

What does all this mean for the global ocean?

It's a mixed signal, but the fact that it's not all negative is itself encouraging. China has forward-looking energy policies, but it is determined to grow quickly. For example, half the buildings constructed in the world from now to 2015 are projected to be in China. That will mean greater use of fuel—mainly coal—and consequently the release of more carbon dioxide into the atmosphere and the ocean.

China is a key player in determining the planet's future, and there are reasons to hope that it may be moving in the right direction. But it's not the only country that will tell the tale of the sea. There are also Brazil and Russia, India, with its massive population and fast-paced growth, and, of course, the United States, with its incoherent policies for cutting carbon.

Adding to all of this uncertainty is that we don't know how close we are to a tipping point in either atmospheric or ocean chemistry. Do we have years, decades, centuries, millennia? What is clear though, as Kendall said, is that the global climate and ocean system is changing much more quickly than scientists have predicted, and those predictions have mainly ignored the particular stresses on the sea.

How will climate change affect the billions of humans who depend on the sea? It seems to me that the key is not to look at the richest of us—who may be able to buy our way out of starvation or global panic—but, rather, at those who are already desperately poor, tied to a resource that's giving out. I decide to go and speak with some of the people who live on the Tanzanian island of Zanzibar.

Reading the vital signs: Adaptability 9

ZANZIBAR Where the secrets of the sea converge

Here on the sand-swept shores of Zanzibar, the moon tugs at the ocean as it has since time immemorial, bringing two high tides and two low tides each day. And just as the moon pulls inexorably at the sea, so it pulls at the people who live on these islands. Many Zanzibari live by a lunar calendar, in deference to the rhythms of the sea as well as to their Islamic faith. For them, life passes in time marked by the tides and the moon rather than where the sun is in the sky, a timekeeping method that Stone Age humanoids are thought to have developed. The critical switches between months depend on moonsets, lunations, and lunar crescents. Seasons are governed by the monsoons.

It is a primordial rhythm. Even the sea creatures live by it, spawning and migrating according to signals they get from the moon, in turn triggering the Zanzibari fishermen to launch their wind-powered fishing boats into the Indian Ocean and to lay their rudimentary nets. On the surface it seems that little has changed over the millennia here on this Tanzanian archipelago a few kilometers off the coast of Africa, but deep down global changes have already affected the lives these Zanzibari are leading.

This is the final spot where I see all the secrets of the changing sea converge. Here, in one of the poorest countries on earth, is the place where scientific theory, climate models, acid tests, plankton research, massive swaths of oxygen-starved water, crumbling coral reefs, and a pillaged fishery are already made manifest in the precarious body weight of the Zanzibari and their children. Zanzibar is a hint of how the changing vital signs of the sea will force humans to adapt. It is a sentinel archipelago; the people here on these moon-kissed islands are on the front lines. They have few reserves—either bodily or financial—to bring to the battle. And this is in the relatively early days of ocean stress: Before the sea grows more acidic, which could have catastrophic effects on Zanzibar's remaining coral reefs and the fish that depend on them. Before the rise in sea level, which could devastate villages and drown more corals. Before changes in the sea's surface temperature bleach more of the corals, push the fish toward the cooler water near the south pole, and maybe alter the currents and monsoons that govern life on these islands.

*

Narriman Jiddawi is one of the few Zanzibari to understand the juggernaut that is on its way and to try to construct a defense against it. A fisheries biologist and senior research fellow at Tanzania's Institute of Marine Sciences on Zanzibar's main island—called either Zanzibar Island or Unguja—she was recently named a fellow at the prestigious U.S.-based Pew Institute for Ocean Science for her pioneering work in marine conservation.

Right now she is barefoot in a rental van with a pink linoleum floor, swathed head to ankle in traditional Muslim dress, zooming over potholed roads in the sick-making heat of midday. We are headed toward the Zanzibari villages of Fumba and Bweleo, where Jiddawi plans to show me two of her prized ocean projects. As always she is carrying on several vehement conversations at once: with me, with two graduate students, and with a government fisheries fellow who's come along for the ride.

Suddenly she belts out an order in Kiswahili. The van screeches to a halt. It's early February, by the Julian calendar, and the mangoes will never be better than right now. Some villagers have collected the fallen and are selling them by the side of the road. Out Jiddawi gets, still barefoot, and starts a noisy, imperious negotiation, searching for the biggest, the juiciest, the most perfect mangoes. Triumphant, she returns to the van. She has scored six melon-size mangoes for the equivalent of about 50 cents, and a 20-liter bucket of small mangoes for the same amount. They were charging too little, she tells me, as we take off again. She shakes her head—they don't know what things are worth. She paid them double.

Jiddawi cannot bear the inequities that fall to people, especially those who are poor, African, Muslim, and female. Her professional life is driven by the urge to right these wrongs, and her international reputation is based on the fact that she is capable of dreaming up practical solutions that will help. She is an extraordinarily tenacious woman.

Yesterday at her office in ancient Stone Town, the main city in Zanzibar, she told me that her projects in Bweleo and Fumba fall into the category of applied biology, verging on sociology. Fishing makes up about 10 percent of the gross domestic product of Zanzibar; roughly a quarter of the islanders fish for a living. Almost all of it is artisanal, happening within about 20 meters of shore in boats with no motors. As much as 70 percent of the animal protein that Tanzanians—including the Zanzibari—consume comes from fish. Fishing defines the island and its culture.

However, women are not allowed to fish. If a woman even happens to touch a fishing net, the net is considered cursed. That means that if a husband has a bad fishing day or week or month, the wife can't jump in a boat and try her own luck. She is shut out of the island's main economy and its main mechanism for getting food for her own table.

The coral reefs that fringe the island and provide living space for most of the fish are being systematically destroyed here, as in so many other places. Higher temperatures from global climate change bleach the coral animals and kill them. Fishermen use dynamite or poison to catch their fish. And others kill the coral and dig up their bony structures to make lime for Zanzibar's building industry.

At the same time, fish are becoming so much rarer elsewhere in the ocean that poachers have begun targeting the waters surrounding Zanzibar. If that weren't enough stress on the local ecosystem, the government—eager to alleviate the poverty that gives Tanzania the rank of 159 out of 177 countries on the United Nation's well-being index—opted several years ago to sell deep-sea fishery licenses to foreigners who use the infamous industrial fishing methods of trawling and long lines. In the prophetic analysis of Boris Worm and Ransom Myers, this means that the fish numbers are poised to drop catastrophically in the next couple of years. Together these trends mean that there are fewer places for fish and their young to live—leading to fewer fish, as well as far more pressure to catch those that remain.

On Zanzibar the fallout is that the men are catching fewer fish. And the old men say that the fish they do catch are smaller. This coincides with a dramatic rise in the human population of Zanzibar, up from about 300,000 in the 1970s to around one million now. Additionally, over the past decade or so about 100,000 tourists a year have flocked to Zanzibar's snow-white beaches. Most of them like to eat fish—and not the little fish close to the bottom of the marine food web, but the increasingly rare fish at the top, such as marlin and tuna.

All these factors bring Jiddawi bumping over the axle-breaking roads on her way to Bweleo and Fumba. While women are banned from fishing, over the ages they have been allowed to collect seaweed and shellfish such as oysters and clams from the tidal zones. But as their families have become hungrier and the sea more degraded, the women have begun collecting too many shellfish. So a couple of years ago Jiddawi, with help from Israeli and American scientists, set up some experimental farms in tidal pools where the women can rear oysters, clams, and other shellfish, and either eat them or sell them to tourists. This was an add-on to the low-tech seaweed farms some of the women were already running.

Hauke Kite-Powell, an aquaculture specialist at Woods Hole Oceanographic Institution in the United States who was involved in setting up the farms, has been telling me about them. They started about 1989, when some Zanzibari women began farming seaweed for the international pharmaceutical industry, one of the few obvious inroads the globalized economy has made into this island. The women string lines between stakes in the tidal pools near their villages, tie bits of seaweed to them, let them grow, and then harvest, dry, and sell the seaweed to the pharmaceutical companies, which derive the natural emulsifier carrageenan from it. At first, says Kite-Powell, the women made money. Then the practice of farming seaweed spread, the supply rose, and prices went down drastically. Some of the women at Bweleo still have tiny seaweed farms, but Jiddawi could see the writing on the wall and figured that shellfish farms would help increase the women's income—or, at least, their protein levels.

By the time we get to Bweleo it's a low tide, at the end of a spring tide. In the language of Zanzibar, that means it's the new moon. Four of the several dozen women involved in the projects are waiting for us, barefoot and dressed in colorful hijab: Amina Khamis, Fatma Ramadhan, Subira Vuai, and Mwatum Juma. They are some of the original seaweed farmers who have now added shellfish to their repertoire. They lead us to the seemingly endless beach: pure white sand framed by a sky so blue that most Westerners could not even imagine the color. A few dhows float in the distance. A man tinkers with his nets on the beach, and some boys play with handmade sailboats. It is violently hot under the midday sun.

This is the way most humans picture the ocean, as a relatively simple, flat entity that fills in the spaces between land. It is an ancient, iconic image. I think of those first humans to settle by the sea 150,000 years ago on the South African cape. This must be similar to what they saw. The ocean looks so immutable, so dull—so unimportant.

I can taste salt in the air as I walk with the women toward the sea. The sand sucks me down, footstep by footstep. The sound of the wind fills my ears, to the exclusion of all else. Around me are expansive beds of sea grasses, the most extensive I've ever seen. I think about Heike Lotze and the

Wadden Sea, where the locals destroyed all the sea grasses in the Middle Ages. One of the women bends down from the waist, her skirt dragging in the water, head carefully draped, to show me a sea cucumber and a sea urchin. The others gather around as Jiddawi translates from Kiswahili into English for me, patiently drawing pictures of the creatures in my notebook and writing the Latin names beside.

The women fan out along the beach, each heading toward her own farm. Each farm is a circle of sticks perhaps a couple of meters across. Here an oyster is attached to the formation, there a mussel. The shellfish feed on plankton, so the tides bring them food. The women just give them a place to grow, tend them, and harvest them when they're ready. There's little, if any, interference with the natural systems here. This is not the large-scale, industrial farming of Scotland, China, and North America that feeds big, carnivorous fish with protein pellets made from little, ground-up wild fish and krill, and adds antibiotics and chemical colorings to the pellets, affecting the life below the fish pens. It's a different paradigm.

It also doesn't require much financial outlay from the women or their families. Jiddawi helped the women collect seed stock for the shellfish by getting them a boat so they could go further out to sea to some tiny, less depleted islands. These farms are actually helping the environment by taking pressure off the wild shellfish. Plus, they're bringing the women food.

These women are used to eating little. A study from 2001 showed that more than a third of Zanzibari children are underweight. A more recent look at infants nine to eighteen months old on the island of Pemba found that 81.4 percent were either stunted, wasted, or underweight. Nearly 84 percent were anemic. The life expectancy across the nation of Tanzania is fifty-one years.

Jiddawi explains to me that during the spring tides, known here as *bamvua*, the moon pulls hard at the water, causing especially high and low tides twice each month, for a week at the new moon and again at the full moon. This is when the men fish and, therefore, these families eat. In between the *bamvua* are the neap tides, one week after the new moon and one after the full moon, when the pull of the moon is weaker. There's not so much difference between the low and high tides and the fishermen of Zanzibar are less likely to catch fish. Then, if they have money, they eat. If not, they go hungry.

So when the women of Bweleo started growing seaweed and then rearing shellfish, the men were happy, the women tell me, smiling and nodding. It took some of the pressure off, especially given that catches overall are down. The women are proud of what they've done here. One tells me that she bought a cupboard for her kitchen from the money she made seaweed farming. The cupboard is one of her treasures. Most of all, though, the women are grateful that they can feed their children.

To Kite-Powell this is clearly the way of the future, not just here in this poor community, but the world over. Fish farming is the only way to increase the global supply of seafood, he argues. And humanity has come relatively late to it, medieval carp ponds notwithstanding. "The question is not whether it's possible, but how to go about it," he told me. We have to figure out how to do it right, without fouling the environment, so it can last. Zanzibari shellfish farmers are on the front lines of the effects of ocean change, but just maybe they are also pioneers in a new, low-tech world trend.

*

We passed a smoking structure on the way here, and now one of Jiddawi's graduate students is giving me a chemistry lesson that complements the one Joanie Kleypas gave me about how the seas become more acidic. This lesson also concerns the calcium carbonate that the coral animals use to build their bony skeletons. As the sea becomes more acidic from the carbon dioxide we are putting into the atmosphere, the shell-building calcium becomes less available. This is the most worrisome of all the changes to the ocean because we don't know and can't predict how marine life will react to the new ocean chemistry.

Zanzibar Island is an ancient coral reef. Now dead, the reef became an island during the Pleistocene epoch when the sea level fell. The coral rag, or ancient dead coral, is mined as a key building material on the island. It has been mined for as long as anyone remembers and is one of the island's main industries. Once mined, the rag is put into an open kiln, laid down in layers separated by coconut logs, firewood, and coconut husks, and set to burn for three or four days. When heated, the calcium carbonate of a coral skeleton breaks down into calcium oxide powder—which we know as lime—and carbon dioxide gas.

Builders throughout Zanzibar mix the lime with sand and mud to make cement. They also use the lime for whitewash to paint the exteriors of the island's buildings. It's a key element of the picture-postcard cachet of Stone Town, which is named after the coral stone it's built on. Some builders, however, find it more convenient to make lime out of live coral from the sea than the dead stone under their feet. The practice is endemic and the living reefs are suffering, adding yet another pressure on the coral ecosystem so critical to the Zanzibari—and the world. The practice of making lime is also an excellent way of putting yet more carbon dioxide into the atmosphere.

By the time we get to Fumba, it is too hot to walk on the beach. There's not a cloud in the sky. Jiddawi takes me to see some of the women in the village. We pass an enormous pile of cockle shells, refuse from an old cash crop. The tiny house beside it has walls of shiny lime green instead of the usual whitewash—luxury.

Down at the edge of the beach five of the farming women are waiting for us, sitting in a circle on pieces of cement under a coconut-thatch lean-to, covered head to foot in light drapes of cloth. Jiddawi sits on a rusted barrel stuck in the sand. I still don't understand how the moon—rather than this ever-present, scorching sun—governs their lives. Patiently, they explain that it's not just the moon. It's more complex than that.

The ocean also depends on the winds, they say. Right now, we are in the *kaskazi*, the northeast monsoon season, which runs from December to March by the Julian calendar. Jiddawi sketches the archipelago and the eastern coast of Africa in my notebook in broad, black strokes, drawing in the directions of the winds. During *kaskazi*, the winds come from the north and blast toward the west side of the islands. The other main season, running from June to August, is the *kusi*, or southeast monsoon. Then winds and currents from the south batter the island and blow up against the east coast of Africa. These are stronger than the *kaskazi* winds. The catch depends on the monsoon, as well as the lunar calendar and the tide.

This way of understanding the passing of time is bred in the bones of these women. Time is a rhythm. I hate to imagine how global climate and ocean change will affect these rhythms. Will the monsoons wither or grow? What will the fish do? In a hundred years, will the offspring of these women still count on the moon and the tides and the winds?

The women turn to Jiddawi. Yesterday nine boats were fishing offshore, dragging nets with small mesh. The women have been seeing this more often—another sign of big change. These are illegal fishermen using illegal gear designed to catch fish too young to have reproduced. This whole area—Jiddawi sweeps an arm in the direction of the sea—is a conservation area and shouldn't be fished by outsiders at all.

Back in Stone Town, the sun is mercifully setting. Small boys dressed in white leggings, long shirts, and white caps run merrily through the warren of crumbling streets on their way to prayer with the older men. Their ankles are skinny under their flowing trousers. Mangoes are ripe here too, and the boys toss juicy pits, stripped bare of flesh, into the streets. There are too many to eat, so some of them rot in the streets.

This ancient town is beautiful, redolent of all the historic trends of the sea: the brave quest for trade and spice routes, the dreams of conquerors and migrants, the hardships of slavery, and the fight for emancipation. It is one of the oldest cities in the world and was recently named a World Heritage Site for its distinctive culture and iconic, whitewashed architecture. More than half these architectural jewels are set aside for the poorest of the poor, paid for by the state.

At night tourists congregate on the west side of Stone Town in the grassy garden on the beach. The beauty of the sea is preternatural, the dark of the night absolute. Only the stars give off light, apart from the long row of open-fire barbecues lined up along the beach. Smoke hangs in the air.

Here is the day's catch, cut up and strung on skewers for pennies a stick: marlin, tuna, shark. These are the fish, rich in calories and globally depleted, that the locals never eat, saved for those of us born in wealthier parts of the world. At this moment, the systems we humans have set up seem horribly unfair and intractable.

Finding hope 10

THE DRY TORTUGAS

Journey to the deeps

Hope skitters back and forth across time, lingering both in the past and in the future. Unlike despair, which burns up the present and obviates all else, hope is dependent on context. With hope, either you believe in patterns adduced from the past and extrapolated to the future or you abandon them. In this way, hope is intrinsically absurd; you use the gift of logic, then rely instead on the irrational, the unexpected.

Why do we hope? This is still a mystery, a conundrum that even poets have failed to resolve. We have plausible theories of how the universe came to be, how the genome operates within the cell, how light and time and space function. But we have not parsed the structure of hope, which may be why we long for this remaining mystery to prevail, even as we fear losing faith with it. Or so it seems to me as I prepare to board a tiny submersible vessel that is to take me to the bottom of the sea off the Florida Keys, the final research expedition for this book.

I'm out of hope, mired firmly in the desolate present. I am an empiricist, and the evidence of an advanced illness in the global ocean seems ineluctable. It is as though a cancer whose primary site is the atmosphere has metastasized to the ocean, making the disease far harder to treat. After checking the vital signs, I've begun to think that the ocean is in palliative care and I a mute witness to its death rattle.

Why keep going? I ask myself. *How could another research trip make any difference?* But I had learned of this research from Mike Roman, my tutor on the ScanFish in the Gulf of Mexico, and had made arrangements many months earlier to take this dive 914.4 meters (3,000 feet) to the bottom of the ocean. This was not to be a trip to look at yet another element of the ocean that was going wrong. For the researchers, it was a journey to a part of the ocean never before seen by human eyes, to look for compounds from deep-sea animals that may someday—with ingenuity, patience, and luck—cure cancer. And for me, it was aimed at trying to understand, for a few hours, what it was like to be deep inside the ocean womb, where the human hand had not yet touched. It would be the final test, the end point of my personal journey to conquer the paralyzing fear of being under the surface of the water. It was to be a rite of passage.

Now, however, it seems like a late dash to capture a few last goods from the sea before we spoil them all. It is tinged with irony. Surely hope does not lie at the bottom of the sea.

*

Amy Wright, a marine scientist at Harbor Branch Oceanographic Institution in Florida, is my guide to the ocean floor this June morning. Along with her colleagues Shirley Pomponi and John Reed, she operates the *Johnson-Sea-Link I* and *II*, two of only a couple of dozen crafts in the world that go deep for ocean research. Together, Harbor Branch's twin vessels have logged more than 9,000 trips to the deeps.

All that experience is small comfort as I glimpse the *Johnson-Sea-Link II*, which will carry me below. We only arrived on board its carrier ship, the *Seward Johnson*, late last night and I'm seeing the machine for the first time just as I'm about to get in. The front part, where Wright and the submersible's pilot, Phil Santos, will sit, is a tough, 13-centimeter-thick acrylic sphere loaded with controls. These allow the pilot to drive the submersible once the ship has launched it into the open sea. (That's the difference, I've been told, between a submersible and a submarine; a submersible is dependent on a ship. But while some submersibles are tethered to the mother ship, ours is not. Once it hits the water it's an independent traveler, run off fourteen two-volt battery cells.) The controls also allow the pilot to maneuver a robotic claw, a scoop, and a suction tube to capture whatever deep-sea animals the scientist wants and place them in containers to bring them to the surface.

The back part of the submersible is where I'll be, along with Frank Lombardo, the craft's electronics technician, who is a dead ringer for the American actor George Clooney. It's made of aluminum and gets very cold in the deeps. I've been told to bring everything warm that I own—that's five layers and a blanket.

It dawns on me that at this moment I'm past hope or despair: I'm tight in the grip of fear. Almost everyone else on board is pumped—apart from the nervous grad students who, like me, are slated for their inaugural dives during this week-long expedition—eager to get going, excited to be exploring this part of the ocean, anticipating the treasures yet to be discovered.

Pomponi strolls over and asks me how I'm doing. I tell her I'm scared. She is an eminent scientist and brings a scholar's mind to this information, assembling her facts logically. To her, it's an issue of mathematical probability. "Well," she says, "we've only lost two people in more than 9,000 dives, and that was a long time ago. Both died in the same back chamber you'll be in."

I am silent.

She continues. "The air in your chamber is recycled. A scrubber system removes the carbon dioxide you exhale and puts it into the ocean just outside the chamber. Then we add more oxygen to the air in the chamber to keep you alive. The others died because the submersible got caught and the scrubber stopped working. The carbon dioxide hit lethal levels. But," she assures me, "we've replaced the scrubber since. It's quite safe, quite safe."

With that, I am at the base of the submersible's back entrance, climbing through a hatch into the aluminum chamber. I have to lie down to fit. Lombardo comes up after me, clangs the hatch shut, and lies down beside me. I have the sensation of being shut into the coffin space left when you put two bathtubs one on top of the other. I can sit up only if I hunch my back. We will be in here for a little more than three hours, most of that at around the crushing depth of 3,000 feet.

Lombardo points behind me to the fan for the carbon dioxide scrubber and explains that it will hold the carbon dioxide level of the air inside this chamber at 0.4 percent—as long as I don't block it too much. I press my body away from the fan, acutely aware of the need to avoid interfering with its work. I've done vast research on the composition of the atmosphere, on how the planet's life-support systems work, on the boundaries that limit life, the utter dependence humans have on the planet's complex biological systems, but nothing really brings it home like being here in this microcosmic aluminum chamber. The chemistry and physics of the global ocean are about to become unavoidably real.

There's a gauge that measures the oxygen content of the chamber's air; it needs to stay below about 21 percent. Too much oxygen is as fatal as too little. As well as the gauges, which can fail, there's a more immediate test of how Lombardo and I are feeling. If we start to get dreadful headaches, we'll know that the proportion of gases is off.

Lombardo takes me through a lesson on the emergency gear. There's a backup battery in case the submersible drops its main battery pack. Our chamber's atmosphere is separate from that of the front sphere, which means that if we have problems, the people in the front can still survive, or vice versa. There's a fire extinguisher. (I had failed to worry about the possibility of an oxygen-sucking fire in the chamber until this very moment.) And there is scuba gear on the off chance we need to use its breathing apparatus or enter the water. Of course, scuba divers are not known to thrive below 54 meters and we are going to 915. Still, I figure, it's something.

Then it's time for a primer on what I would need to do to get the submersible to the surface from this back chamber, in case the other three people on board couldn't do it. I am terribly alert as Lombardo talks.

Then we are off, launched from the ship by a contraption that reminds me of a giant swing set. We descend at the rapid pace of 100 feet a minute into this mystery-filled third dimension of the ocean. Lombardo checks the seal on the hatch to make sure no water is pouring in (another risk I had failed to worry about, and it occurs to me that there must be many more) and settles down to read a paperback novel in the dim light.

To him, to Wright and Santos, to Pomponi and Reed, this is all in a day's work, the way flying is for airline crew. Each has been down hundreds of times. Lombardo stopped counting a couple of years ago when he got to 650 dives. The thrill for them is the quest to find new deep-sea animals that may produce unheard-of compounds that could help medicine.

Why deep-sea animals? The deep-sea sponges, sea fans, and other creatures that they're in search of can't move. In scientific terms they're sessile, meaning they can't stroll over to a potential mate and tap him on the shoulder. They can't leap across a breach to capture a tasty bite of lunch or run to hide from a hungry enemy. To compensate, these stationary creatures of the deeps have evolved a sophisticated arsenal of chemicals—unknown in the rest of nature and unimagined by humans—that allow them to fend off predators, attract prey, and lure mates.

The search for drugs from the sea got under way in 1967, when the National Institutes of Health in the United States began to sponsor research, and Wright is one of the field's international superstars. Her chemical research has been a key, for example, in the development of the drug Yondelis (also known as ecteinascidin 743, or ET-743). If it's approved, it will be the first drug derived from a marine creature to be used on cancer patients. As we descend today, Yondelis is in the third phase of trials in Europe for use as a drug of last resort for ovarian cancer. It's showing promise in treating metastasized cancers of the breast, colon, and lung, as well as skin cancer. Wright is crossing her fingers that it will be approved for use in Europe soon.

In 1985 and 1986 she went collecting in the Florida Keys and found a little sea squirt—*Ecteinascidia turbinata*—in a marine mangrove forest. Her team ground it up and did a trial on its chemical compounds, only to find that it worked incredibly well against leukemia in animals. Wright then purified the compound. Some of its chemical structure had already been worked out by her colleague Ken Rinehart in Illinois, but Wright pinpointed the final piece, a process very like solving a three-dimensional jigsaw puzzle. The drug that's now in trials is a synthetic version of the same compound, not extracted from live sea squirts.

The fascinating thing about Yondelis is that it works on cancer cells in a way that is different from other cancer drugs. It binds with the minor groove in the helical structure of the cell's DNA and triggers apoptosis, making the cell kill itself. One of the characteristics of cancer cells in general is that they fail to self-destruct when they ought to, but instead keep reproducing. Triggering apoptosis in a cancer cell—so that the healthy cells can survive and flourish—is one of the dreams of modern cancer research. The compound in Yondelis also appears to interfere with a membrane protein that protects the cancerous cell from absorbing toxic substances, such as chemotherapy drugs. In other words, the drug breaks down the cancer cell's dogged ability to stay alive, and this is why doctors have high hopes for its use as part of an anticancer cocktail. In particular, Yondelis is the first new drug in more than twenty years to treat the rare but deadly class of soft-tissue sarcomas.

I can hear Wright now, narrating a video that is being made of everything she's seeing. A monitor measuring about 13 centimeters square is above my head, allowing me to track what she sees from her seat in the front bubble. I also have my own view into the deeps through two tiny, thick portholes. About two minutes into the dive, at about 60 meters, we lose sight of the sun's rays. It is, as the abyssal explorer Robert Ballard writes in his book *The Eternal Darkness: A Personal History of Deep-Sea Exploration*, "much more oppressive than the blackest chamber inside any cavern on land."

Suddenly, I think of Olaus Magnus and the map he made of the North Sea in 1539. For the first time I understand exactly why his ocean was thick with fearsome monsters. It's scary and alien down here. It is like facing the demons within.

As we descend tiny luminescent organisms flicker and flash outside, the only points of light. We see marine snow, clumps of decomposing bodies—mainly plankton—that fall to the floor like flakes of snow to be recycled as food and organic chemicals for other creatures. We are witnessing carbon being stored in the deep ocean, where it will stay for centuries or more, depending on where it falls. This is the day-to-day working of the complex chemistry of life on earth.

It's getting colder.

Santos announces that we're at 2,000 feet now. I'm shivering, wearing every jacket I've brought on board. Lombardo, reading his paperback, casually pulls up the hood of his sweatshirt. The monitor says it's 6.9 degrees Celsius; up at the surface it was something like 30 degrees Celsius. Light doesn't penetrate this far, and neither does heat. Nor, for the most part, do humans. Most of those who do come are specialist scientists and naval crew.

All I can think about is the deadly weight of the water on top of me, and the fact that the water's physical imperative is to get into this nice, empty space Lombardo and I are inhabiting. The physics work like this: at the surface of the ocean, you bear the weight of one atmosphere of pressure, or about 15 pounds per square inch. Our bodies are perfectly suited to that. For every 33 feet you go below the surface of the ocean, you add the pressure of another full atmosphere. That brings the total to 30 pounds per square inch at a depth of 33 feet, or 90 pounds per square inch at 165 feet, close to the limit for scuba divers. If you're a deep snorkeler, a free-diver, or a scuba diver, you know the feeling of the water's pressure on your eardrums as you descend. Here at 2,000 feet below the surface—and dropping—the pressure is already the equivalent of more than 60 atmospheres, or 900 pounds per square inch. Once we get to the bottom, at 3,000 feet, it will be more than 90 atmospheres, or more than 1,350 pounds per square inch.

As Ballard explains, the animals that live in the depths are adapted to the pressure because they are filled with water from the same depth. "A diving craft, however, is a hollow chamber, rudely displacing the water around it," he writes in *The Eternal Darkness*, a passage that is burned indelibly in my memory as I lie here in this frigid shell. "That chamber must withstand the full brunt of deep-sea pressure. . . . If seawater with that much pressure behind it ever finds a way to break inside, it explodes through the hole with laserlike intensity. A human body would be sliced in two by a sheet of invading water, or drilled clean through by a narrow (even a pinhole) stream, or crushed to a shapeless blob by a total implosion."

This is by no means the deepest slice of the ocean floor. Like the land surface of the planet—except more extreme on all counts—the ocean's bottom has high peaks, medium-size hills, and low valleys. The very deepest part is the Mariana Trench in the Pacific Ocean, which descends to nearly 36,000 feet below sea level. Only two humans have seen it, and only once, on January 23, 1960. But even at 3,000 feet a leak or a crack in our hull would be instantly fatal. This is one reason the acrylic front chamber and the aluminum back chamber are separate. If one is damaged, the other might still remain intact and be able to return to the surface.

The immense weight of ocean water has been one of the big barriers to deep-sea exploration. Not only does it make deep diving dangerous, but devising hulls that can withstand that pressure also makes it expensive. Humans have spent centuries trying to get past the first few meters in a range of diving bells, bathyspheres, and other contraptions. It was only in 1930 that William Beebe and Otis Barton became the first humans to descend far under the surface of the ocean. Their submersible reached a depth of 1,426 feet but stayed for mere moments. As Ballard recounts, Beebe said that looking down into the deeps was like seeing "the black pit-mouth of hell itself." Four years later the same pair reached 3,028 feet, remaining there for three minutes. That record for depth stayed intact for fourteen years.

Since then humans have figured out how to go deeper. This craft I'm on, the *Johnson-Sea-Link II*, can go to 3,000 feet. The more famous *Alvin*, a submersible run by the Woods Hole Oceanographic Institution in Massachusetts and owned by the U.S. Navy (Ballard used it to find the *Titanic*), can reach nearly 15,000 feet and stay down for as long as ten hours at a time.

Rather too casually, I ask Lombardo what would happen if we sprang a leak. "I would fix it," he says calmly. He says he's not worried about "catastrophic" leaks because he and all the other nonscientists who travel on the submersible also do the repairs and maintenance on it. They're highly motivated to make sure things work well. Still, the feeling of vulnerability is a thing alive. We four representatives of our voracious, casually destructive species are here in the belly of the planet's fundamental life force, at the mercy of a system we are only beginning to understand. We are immersed. There's no way out except back up the way we came.

Then we hit bottom: 3,000 feet. It's now 5.9 degrees Celsius. The cold is viscous.

Santos turns on the vessel's xenon arc lights, which illuminate the inky marine blackness for about 50 feet in front of us. We are at the bottom of a sheer-faced sinkhole in the Tortugas Valley, beyond the continental shelf south of Florida. The wall in front of us runs straight up for 390 feet, stark as an iceberg. The scientists are hoping that this kind of wall will be prime real estate for deep-sea sponges and sea fans, which is why we're here. Wright is on the lookout for specimens to take to the surface.

Lombardo has abandoned his novel. "It's pretty cool," he murmurs, peering out the porthole. "Very few people have ever been this deep. Nobody has ever seen this before." We are the first, and perhaps the last.

I see hollows and crevices and animal boreholes in the wall, and little shrimp poking their heads out. Wright says, with disgust in her voice, that the wall face looks as though it's made of Silly Putty. She was hoping for a hard rock surface because her sponges like to attach to that.

Life is everywhere. Here, a field of tiny pink sea fans, an anemone, a pile of coral rubble; there, eels, unidentified fish, a giant squid, crabs, a black lobster with eyes that glow eerily white, some ancient brachiopods whose ancestors first appeared more than three billion years ago. A small fish swims past my head as I peer out the porthole; it is no bigger than my fingernail, delicate pink. All around us is the marine snow, suspended as if in solution.

The hand of *Homo sapiens* is here but invisible. Even this deep, the fossil fuel–based carbon dioxide that we have pumped into the atmosphere can be tracked chemically. At this depth, salinity patterns have already been altered by the volume of glacier ice melting as the air heats up, and the temperature of this water has also risen slightly.

Reed—who has boxes of decades-old, carefully annotated geological papers back on the ship for his dive-site research—has explained to me that humans only became aware of the topography of this part of the ocean in 1838, during the first hydrographic survey of the area. More than a century later, in 1954, a seismic survey running from Key West to Cuba discovered this sinkhole, an unusual ocean-floor formation. In the 1990s a nuclear submarine passed over the area with a side-scan sonar to catch a clearer outline of the sinkhole's shape. But by the time we, the first people, arrived for the first time, this unexplored, unexploited corner of the ocean had already been chemically altered by human activity elsewhere.

I think about a dinner I had a few months ago in England, with the brilliant scientist Carol Turley of the Plymouth Marine Laboratory. She's known as the acid queen in the UK for her formidable work explaining the changing pH of the global ocean. She's a big thinker, a synthesizer, like Tim Flannery. I've come to think of her as the oracle of the deeps.

Some of her earlier work was on marine snow, and she spent part of our dinner explaining the characteristics of the planet's deep-ocean water to me. Like the rest of the planet's deep water, the batch our submersible is in right now is made up of water of different ages, created in all sorts of different places at different times. It is forever in motion, but different parts of it are moving in different directions. As we forage around down here, we cut across the dimension of time, through past, present, and future water. And there's another layer of time laid overtop of this. Some of the biological processes of this part of the ocean are local and immediate—like the plankton blooms at the surface, the day-to-day activity of fish and other marine creatures. Some are tied to the ocean's seasonal cycles, some to annual patterns, and some to events that that happen only once in several million years.

It's like being immersed in the Australian Aboriginal concept of Dreamtime, or anywhen, when time ceases being a straight line or even a cycle, and all time exists simultaneously. Previously I have understood this idea only intellectually. Now, I feel it. I live it. It is indescribably shattering.

People who go to extreme environments—astronauts and deep divers—often report odd experiences. It has something to do with being pushed past a limit. Some become emotional, focusing on intensely personal issues such as an unresolved relationship with a parent. Others turn to a religious faith. Yet others go on a surprising intellectual journey to the abstract. This is where I have gone, deep into the theories of time and dimension. And now the epiphany comes. My fear vanishes, as does my despair. I feel as though I have burst through this cramped space and out into the whole world.

I'm flooded with hope, with a sweet consciousness of the rich march of time, stretching from the deep past to the remote future, each moment containing all the others. It is impossible to think only in the self-indulgent, despairing, fearful present when surrounded by life across these four dimensions. All this time, through all these voyages, I've been trying to reason my way to hope, to convince myself that hope is justified, to build a case. It's been the question I asked of all the scientists I met.

In fact, hope just is. You can't run through a checklist to get to it. It is absurd and irrational. But, like love, it is human. Like laughter, hope catches and spreads. It expands logarithmically, like the changes now under way on our planet, like our growing understanding of them, and like our powerful collective human ability to start coping with them.

That doesn't mean that hope is naïve. Shivering in my undersea womb, peering at these wondrous, ancient life forms, it occurs to me that we are in an era that holds out the potential of magnificent regeneration. We could, if enough of us wanted to, form a new relationship with our planet. We could become the gentle symbionts we were meant to be instead of the planetary parasites we have unwittingly become. Perhaps this is the system switch that will be in the offing. Instead of sending the ocean lurching further into an irrevocably altered state, maybe humans will irrevocably alter our relationship to it and understand that we must keep it healthy if we are to save ourselves.

The point, as Turley explained to me, is that biology—and that includes our species—is at work all the time. It is flexible, adaptable. It works in four dimensions all at once: breadth, width, depth, and time. It can always surprise us, can always do the unexpected. This is the ocean's prophecy.

Wright has spotted her sponges. They are otherworldly. One is yellow. A second is enormous, shaped like a white glass tulip on a reddish stalk almost a meter high. The tulip-shaped sponge has long silica fibers attaching it fiercely to the sandy bottom. Clear and long, these tenacious organisms are the original fiber-optic material. Santos collects them carefully with the robotic implements and puts them in containers hooked to the front of the submersible, so we can take them to the lab on board the ship hovering so far above us.

This isn't as good a haul as Wright wanted, but it's new genetic material to work on. Maybe these animals will be the ones to reveal a surprising new way to nudge cancer cells toward apoptosis.

Now it's time to wrap up. The batteries are getting low; we're at our limit. Up we go, back to the surface.

*

Three months after we finish our trip, the European Commission approves Yondelis for use throughout Europe against soft-tissue sarcomas that can't be cured by anything else.

Epilogue: A call for wisdom

Truth lies in the tales we tell rather than in the scientific facts that give rise to them. So, understanding the meaning of the changes to the global ocean requires more than knowing that human activity is pushing the ocean's chemistry and biology toward a state it has not been in for millions of years. It takes stepping outside the science and into the curious realm of human psychology, where the stories we weave around the facts are born.

It is possible, for example, to look at what's happening to the global ocean and conclude that it is insignificant, either because the science must be wrong or because humans simply can't have this much power over planetary systems.

It is possible to look at the facts and decide that the ocean will right itself without our help, possibly through an outside force swooping in, whether divine, technological, or otherwise superhuman. Faced with croplands parched by climate change, some pray. This is one of humanity's most ancient and cherished stories, and it runs through culture and time. A key underpinning of this tale is the notion that humans deserve salvation.

Or it is possible to be overwhelmed with guilt, to conclude that we may as well carry on as we are because we deserve to be annihilated for being such idiots. Or by hopelessness: the changes to the global ocean are already so serious that we're doomed, so we may as well dance the merry jig of death. This is the mentality that spread across Europe during the plague years of the fourteenth century.

I'd like to suggest a different story: ocean change is extremely serious, but we have some power to halt or reverse it if we alter our actions rapidly, profoundly, and en masse. In the archetypal human narrative, we face our demons, plunging into the underworld or the psychological abyss, wrestling with the devil's temptations, and returning to tell the tale. We triumph, we lead, we heal. This is Odysseus, Aeneas, Hercules, Sinbad. It is the human epic.

The story we tell matters because it alone determines the actions we take or fail to take. In other words, the final vital sign, the one that will fix the fate of the global ocean, is how the agent of destruction—us—reacts. Will we turn the destruction off? Will we bring on our own destruction so that the earth can survive? Will we continue to attack the organism of the earth, pushing it into a new system that will be unlikely to harbor us?

The problem of the atmosphere and the ocean is a problem of human behavior. As Manuel Barange, director of the Global Ocean Ecosystem Dynamics program based in Plymouth, puts it, humans have unwittingly become an intrinsic part of the marine system. The future is in our hands. It strikes me as critically important that we understand this.

Monica Sharma, a physician who works with the United Nations, delves deep into the human psyche to shift behavior on seemingly intractable problems such as the practice of female genital mutilation and the spread of HIV/AIDS. I ran into her in New York recently. Her view is that for transformation to happen, we first need to understand that transformation is possible. And for that to happen, we need to strip ourselves psychologically naked and figure out what each of us stands for. What is the story about the world that makes sense to us emotionally? What is it that we believe? What are we here for? Once we know that, we can start to ask the right questions, including: What's missing? Answering that question leads to a course of action.

When Sharma sits down with religious and community leaders whose culture tells them that girls must be genitally mutilated, she doesn't judge them or tell them that they're wrong or tell them what to do. She asks them what they stand for. It takes days of deep discussion. It is very hard work. Once they can articulate where they stand—what their story is—she invites them to start asking questions, such as whether a practice such as female mutilation fits. If not, then why are they insisting it be done? She gives them the great gift of seeing the space between what they are and what they do. In many cases these leaders go back to their communities and stop the harm to their children.

A key to her success is that she doesn't give a prescription. Instead, she helps unleash imagination and creativity. In Africa, communities she and others have worked with have come up with more than 5,000 different ways of combating the spread of HIV/AIDS. The power is in the lack of dogma, in starting with the story, in honoring confusion, discomfort, and fear. Sharma refers to this as a call for wisdom.

Here's my question, at the end of this long journey. *What do you stand for? What story do you tell yourself about why you're here?* Perhaps you are a hero. It's clear the world needs heroes right now to ask what's missing and how that can be made right. I don't know exactly how the heroes will gather and how they will alter the course of the planet—nobody does at this point, except that it will have to do with carbon dioxide concentrations in the atmosphere—but I know that if enough people are asking the right questions, we can make a start. This is a call for wisdom, not for logic; for hope rather than despair. It is about taking a stand and then acting on it—being fully human at a time when we need it most.

There's another piece to this. Many of the scientists I interviewed set the drop-dead point for effective action to halt the planet's slide toward chaos somewhere between 2015 and 2030. Others said emphatically that if the concentration of carbon dioxide in the atmosphere rises above 450 parts per million by volume, that will represent a point of no return. Today, it is 387 and rising faster that at any time since humans appeared on the planet.

If you believe that this matters and that something can be done, then the rest of the story reads that the time to act is right now.

Acknowledgments

My most profound thanks to all the scientists who spent so much time explaining their research and reviewing several of the chapters for me, including two anonymous reviewers for the University of Chicago Press. You are an incredibly patient and gracious lot. Some of you even allowed me into your homes and your personal thoughts. I am grateful. I want to make it clear, though, that I alone am responsible for any errors of fact or interpretation here.

In particular, I'd like to thank Nancy Knowlton. You were my starting place, and everything and everyone I encountered after seemed to reach back to you. Dave Robins, you are an unparalleled host and guide and I thank you for all the doors you opened for me in England. Carol Turley, you literally changed my way of seeing over our dinner in Plymouth. I honor you as a remarkable human being. Thanks as well for the extraordinary kindness of Joanie Kleypas, Jim Hendee (and your coral list), Chris Langdon, Mike Roman, Stu Ludsin, Katharina Fabricius, Michael Depledge, Heike Lotze, Boris Worm, Matt Huber, Narriman Jiddawi, Amy Wright, Shirley Pomponi, Phil and Daphne Hart, and many others.

Arthur Hanson provided valuable insight and feedback on the China chapter. Trevor Platt was unfailingly generous on the tricky subjects of plankton and ocean metabolism. And Tim Flannery was a recurring source of support and advice. Thank you all.

My deepest gratitude to Neil Young. Four of your albums provided the soundtrack to my travels for this book, and were often the encouragement to keep going: *Greendale*, *Prairie Wind*, *Living with War*, and *Live at Massey Hall 1971*.

In my personal life, I'd like to thank Sally Harding, my agent. You are this book's midwife. Thanks to Ross Mitchell for your insight on the dimension of time. Many thanks to the redoubtable Patterson clan for the endless challenge and support you provide. David and Anne have been unstintingly generous with ideas, particularly on systems theory, and other much-needed resources. John and Thea, you read very early versions of the first part of the book and offered valuable insight, as well as a carbon-neutral cabin in the wintry Canadian woods in which to write. And without the practical and intellectual gifts of my dear James, this book would never have come to be. Thank you.

Selected bibliography

General references

Ballard, Robert D., and Will Hively. *The Eternal Darkness: A Personal History of Deep-Sea Exploration*. Princeton: Princeton University Press, 2000.
Ballard, Robert D., and Michael Sweeney. *Return to Titanic*. Washington DC: National Geographic, 2004.
Borgese, Elisabeth Mann. *The Oceanic Circle*. Tokyo: United Nations University Press, 1999.
Burnie, David. *The Concise Encyclopedia of the Human Body*. New York: Dorling Kindersley, 1997.
———. *Dictionary of Nature*. Vancouver: Raincoast Books, 1994.
Cahill, Thomas. *Mysteries of the Middle Ages: The Rise of Feminism, Science, and Art from the Cults of Catholic Europe*. New York: Nan A. Talese, 2006.
Carson, Rachel. *The Sea around Us*. London: Oxford University Press, 1951.
Chatwin, Bruce. *Songlines*. London: Random House UK, 2008.
Cousteau, Jean-Michel. *Mon père le commandant*. Paris: L'Archipel, 2004.
Darwin, Charles. *Charles Darwin's Beagle Diary*. New York: Cambridge University Press, 2001.
———. *The Origin of Species by Means of Natural Selection: The Preservation of Favoured Races in the Struggle for Life*. London: Penguin Classics, 1982.
Denny, Mark. *Air and Water: The Biology and Physics of Life's Media*. Princeton: Princeton University Press, 1995.
Ellis, Richard. *Aquagenesis: The Origin and Evolution of Life in the Sea*. Boston: Penguin, 2003.
Estes, James. *Whales, Whaling, and Ocean Ecosystems*. Berkeley: University of California Press, 2007.
Feynman, Richard P. *The Pleasure of Finding Things Out: The Best Short Works of Richard P. Feynman*. New York: Basic Books, 2000.
Feynman, Richard P., and Matthew L. Sands. *Six Easy Pieces: Essentials of Physics, Explained by Its Most Brilliant Teacher*. New York: Perseus Books Group, 1996.
Flannery, Tim. *Weather Makers: How We Are Changing the Climate and What It Means for Life on Earth*. Sydney: Allen Lane, 2006.
Francis, Clare, and Warren Tute. *The Commanding Sea*. London : British Broadcasting Corporation/Pelham Books, 1981.

German Advisory Council on Global Change. *The Future Oceans—Warming Up, Rising High, Turning Sour: Special Report*. Berlin: German Advisory Council on Global Change Secretariat, 2006.

Global Forum on Oceans, Coasts, and Islands. Fourth Global Conference on Oceans, Coasts, and Islands, "Advancing Ecosystem Management and Integrated Coastal and Ocean Management in the Context of Climate Change" (April 7–11, 2008, Hanoi, Vietnam). http://www.globaloceans.org/globalconferences/2008/index.html (accessed May 24, 2008).

Gore, Al. *An Inconvenient Truth: The Planetary Emergency of Global Warming and What We Can Do about It*. Emmaus, PA: Rodale Books, 2006.

Holm, Poul. *The Exploited Seas: New Directions for Marine Environmental History*. St. John's, NL: Census of Marine Life, 2001.

Hume, Stephen Eaton, Betty C. Keller, Otto Langer, Rosella M. Leslie, Alexandra Morton, and Don Staniford. *A Stain upon the Sea: West Coast Salmon Farming*. Madeira Park, BC: Harbour Publishing, 2005.

Huntington, Ellsworth, and Stephen Sargent Visher. *Climatic Changes: Their Nature and Causes*. New Haven: Yale University Press, 1922.

Intergovernmental Panel on Climate Change (Core Writing Team, R. K. Pachauri, and A. Reisinger, eds.). Climate Change 2007: Assessment Report. Geneva: IPCC, 2007. http://www.ipcc.ch/ipccreports/ar4-syr.htm (accessed May 26, 2008).

International Institute for Sustainable Development, Reporting Services. Summary of Fourth Global Conference on Oceans, Coasts, and Islands, "Advancing Ecosystem Management and Integrated Ocean and Coastal Management in the Context of Climate Change" (April 7–11, 2008, Hanoi,—Vietnam). http://www.iisd.ca/ymb/sdoh4/ymb/ymbvol68num4e.html (accessed May 24, 2008).

King, Roger J. B., and Mike W. Robins. *Cancer Biology*. 3rd ed. Harlow, Essex: Pearson Educational Limited, 2006.

Kunzig, Robert. *Mapping the Deep: The Extraordinary Story of Ocean Science*. New York: W. W. Norton, 2000.

Leier, Manfred. *World Atlas of the Oceans*. Toronto: Key Porter Books, 2000.

Levitin, Daniel J. *This Is Your Brain on Music: The Science of a Human Obsession*. New York: Plume, 2007.

Lovelock, James. *The Revenge of Gaia: Why the Earth Is Fighting Back—and How We Can Still Save Humanity*. New York: Allen Lane, 2006.

Maas, Terry, and David Sipperly. *Freedive!* New York: Bluewater Freedivers, 1998.

MacInnis, Joe. *James Cameron's Aliens of the Deep: Voyages to the Strange World of the Deep Ocean*. Washington, DC: National Geographic, 2005.

Magnus, Olaus, and P. G. Foote. *A Description of the Northern Peoples, 1555*. 3 vols. Hakluyt Society, 2nd ser., nos. 182, 187, 188. London: Hakluyt Society, 1996.

McCullough, David. *The Path between the Seas: The Creation of the Panama Canal, 1870–1914*. New York: Simon & Schuster, 1978.

Mitchell, Alanna. *Dancing at the Dead Sea: Tracking the World's Environmental Hotspots*. Chicago: University of Chicago Press, 2005.

Muecke, Stephen, and Adam Shoemaker. *Aboriginal Australians: First Nations of an Ancient Continent*. London: Thames & Hudson, 2004.

Nouvian, Claire. *The Deep: The Extraordinary Creatures of the Abyss*. Chicago: University of Chicago Press, 2007.

Parker, K. Langloh. *Wise Women of the Dreamtime: Aboriginal Tales of the Ancestral Powers*. Rochester, VT: Park Street Press, 1993.

Ponting, Clive. *A Green History of the World: The Environment and the Collapse of Great Civilizations*. Boston: Penguin, 1993.

Reed, A. W. *Aboriginal Myths: Tales of the Dreamtime*. Seattle: Reed Natural History Australia, 1998.

Russell, Dick. *Eye of the Whale: Epic Passage from Baja to Siberia*. New York, NY: Simon & Schuster, 2001.

Short, John Rennie. *The World through Maps: A History of Cartography*. Toronto: Firefly Books, 2003.

Silvertown, Jonathan. *Demons in Eden: The Paradox of Plant Diversity*. Chicago: University of Chicago Press, 2005.

Steffen, W. *Global Change and the Earth System: A Planet under Pressure*. Global Change—the IGBP Series. New York: Springer, 2005.

Stern, Sir Nicholas. "Stern Review on the Economics of Climate Change." http://www.hm-treasury.gov.uk/stern_review_report.htm (accessed May 24, 2008).

Stewart, Iain. *Journeys from the Centre of the Earth: How Geology Shaped Civilization*. London: Century, 2005.

Stow, Dorrik. *Oceans: An Illustrated Reference*. Chicago: University of Chicago Press, 2006.

Suzuki, David, and Amanda McConnell. *The Sacred Balance: Rediscovering Our Place in Nature*. Seattle: Mountaineers Books, 2002.

Tidwell, Mike. *Bayou Farewell: The Rich Life and Tragic Death of Louisiana's Cajun Coast*. New York: Vintage, 2004.

Valiela, Ivan. *Global Coastal Change*. Malden, MA: Wiley-Blackwell, 2006.

Weisman, Alan. *The World without Us*. New York: Thomas Dunne Books, 2007.

Wilson, Edward O. *Consilience: The Unity of Knowledge*. New York: Knopf, 1998.

———. *The Future of Life*. New York: Knopf, 2002.

Young, J. Z. *The Life of Vertebrates*. Oxford: Clarendon Press, 1991.

Papers on coral

Brodie, J., K. Fabricius, G. De'ath, and K. Okaji. "Are Increased Nutrient Inputs Responsible for More Outbreaks of Crown-of-Thorns Starfish? An Appraisal of the Evidence." *Marine Pollution Bulletin* 51 (2005): 266–78.

Bruno, John F., and Elizabeth R. Selig. "Regional Decline of Coral Cover in the Indo-Pacific: Timing, Extent, and Subregional Comparisons." *PLoS ONE* 2 (2007): e711.

"Coral Reefs and Climate Change." Australian Institute of Marine Science Briefing Paper (February 1, 2006).

Doebeli, Michael, and Nancy Knowlton. "The Evolution of Interspecific Mutualisms." *PNAS* 95 (1998): 8676–80.

Donner, Simon D., and David Potere. "The Inequity of the Global Threat to Coral Reefs: Viewpoint." *BioScience* 57 (2007): 214–15.

"Effects of Terrestrial Runoff of Sediments, Nutrients and Other Pollutants on Coral Reefs." International Society for Reef Studies Briefing Paper 3 (2004). 18 pp.

Fabricius, K. E. "Effects of Terrestrial Runoff on the Ecology of Corals and Coral Reefs: Review and Synthesis." *Marine Pollution Bulletin* 50 (2005): 125–46.

Fabricius, Katharina, and Glenn De'ath. "Identifying Ecological Change and Its Causes: A Case Study on Coral Reefs." *Ecological Applications* 14 (2004): 1448–65.

Fukami, Hironobu. "Conventional Taxonomy Obscures Deep Divergence between Pacific and Atlantic Corals." *Nature* 427 (2004): 832–35.

Fukami, Hironobu, Ann F. Budd, Don R. Levitan, Javier Jara, Ralf Kersanach, and Nancy Knowlton. "Geographic Differences in Species Boundaries among Members of the *Montastraea annularis* Complex Based on Molecular and Morphological Markers." *Evolution* 58 (2004): 324–37.

Grimsditch, Gabriel D., and Rodney V. Salm. *Coral Reef Resilience and Resistance to Bleaching*. Gland, Switzerland: World Conservation Union (IUCN), 2005.

Jackson, Jeremy B. C. "Ecological Extinction and Evolution in the Brave New Ocean." *PNAS* 105 (2008): 11458–65

Johnson, J. E., and P. A. Marshall, eds. *Climate Change and the Great Barrier Reef*. Australia: Great Barrier Reef Marine Park Authority and Australian Greenhouse Office, 2007.

Knowlton, Nancy. "The Future of Coral Reefs: Colloquium." *PNAS* 98 (2001): 5419–25.

———. "Multiple 'Stable' States and the Conservation of Marine Ecosystems." *Progress in Oceanography* 60 (2004): 387–96.

———. "Who Are the Players on Coral Reefs and Does It Matter? The Importance of Coral Taxonomy for Coral Reef Management." *Bulletin of Marine Science* 69 (2001): 305–8.

Knowlton, Nancy, Judith C. Lang, M. Christine Ronney, and Patricia Clifford. "Evidence for Delayed Mortality in Hurricane-damaged Jamaican Staghorn Corals." *Nature* 294 (1981): 251–52.

"Last Days of the Ocean." *Mother Jones*, March–April 2006.

Levitan, Don R., Hironobu Fukami, Javier Jara, et al. "Mechanisms of Reproductive Isolation among Sympatric Broadcast-Spawning Corals of the *Montastraea annularis* Species Complex." *Evolution* 58 (2004): 308–23.

Roberts, Santi, and Michael Hirshfield. "Deep-sea Corals: Out of Sight, but no Longer out of Mind." *Frontiers in Ecology and the Environment* 2 (2004): 123–30.

Sala, E., and N. Knowlton. "Global Marine Biodiversity Trends." *Annual Review of the Environment and Resources* 31 (2006): 93–122.

United Nations Environment Programme, Global Programme of Action for the Protection of the Marine Environment from Land-based Activities. "The State of the Marine Environment: Trends and Processes." http://www.gpa.unep.org/documents/soe_-_trends_and_english.pdf (accessed May 25, 2008).

Veron, J. E. N. *A Reef in Time: The Great Barrier Reef from Beginning to End*. Cambridge: Belknap Press, 2008.

Wilkinson, Clive. *Status of Coral Reefs of the World: 2004*. Townsville, Queensland: Australian Institute of Marine Science, 2004.

Papers on hypoxia and dead zones

An Assessment of Coastal Hypoxia and Eutrophication in U.S. Waters. Washington, DC: National Science and Technology Council Committee on Environment and Natural Resources, 2003.

Breitburg, Denise. "Effects of Hypoxia, and the Balance between Hypoxia and Enrichment, on Coastal Fishes and Fisheries." *Estuaries* 14 (2002): 1448–65.

Chan, F., J. A. Barth, J. Lubchenco, H. Weeks, W. T. Peterson, and B. A. Menge. "Emergence of Anoxia in the California Current Large Marine Ecosystem." *Science* 319 (2008): 920.

Diaz, Robert J. "Overview of Hypoxia around the World." *Journal of Environmental Quality* 3 (2001): 275–81.

Diaz, Robert J. and Rutger Rosenberg. "Spreading Dead Zones and Consequences for Marine Ecosystems." *Science* 321 (2008): 926–29.

Ferber, Dan. "Keeping the Stygian Waters at Bay." *Science* 291 (2001): 968–73.

Hotinski, Roberta M., K. L. Bice, L. R. Kump, R. G. Najjar, and M. A. Arthur. "Ocean Stagnation and End-Permian Anoxia." *Geology* 29:1 (2001): 7–10.

Integrated Assessment of Hypoxia in the Northern Gulf of Mexico. Washington, DC: National Science and Technology Council Committee on Environment and Natural Resources, 2000.

Joyce, Stephanie. "The Dead Zones: Oxygen-Starved Coastal Waters." *Environmental Health Perspectives* 108 (2000): 120–25.

Justic, Dubravko, Nancy N. Rabalais, and R. Eugene Turner. "Coupling between Climate Variability and Coastal Eutrophication: Evidence and Outlook for the Northern Gulf of Mexico." *Journal of Sea Research* 54 (2005): 25–35.

———. "Effects of Climate Change on Hypoxia in Coastal Waters: A Doubled CO_2 Scenario for the Northern Gulf of Mexico." *Limnological Oceanography* 41 (1996): 992–1003.

———. "Impacts of Climate Change on Net Productivity of Coastal Waters: Implications for Carbon Budgets and Hypoxia." *Climate Research* 8 (1997): 225–37.

———. "Modeling the Impacts of Decadal Changes in Riverine Nutrient Fluxes on Coastal Eutrophication near the Mississippi River Delta." *Ecological Modelling* 152 (2002): 33–46.

Neretin, L. N. *Past and Present Water Column Anoxia*. Netherlands: Springer, 2006.

Pelley, Janet. "Is Coastal Eutrophication out of Control?" *Environmental Science & Technology* October 1 (1998): 462–66.

Platon, Emil, Barun K. Sen Gupta, Nancy N. Rabalais, and R. Eugene Turner. "Effect of Seasonal Hypoxia on the Benthic Foraminiferal Community of the Louisiana Inner Continental Shelf: The 20th Century Record." *Marine Micropaleontology* 54 (2005): 263–83.

Pocock, John. "Chokehold on Farm Nutrients? Washington Looks to the Nation's Agricultural Midsection to Solve Environmental Degradation in the Gulf of Mexico." *Farmer/Dakota Farmer* September (1999): 14–33.

Rabalais, Nancy N. "Nitrogen in Aquatic Ecosystems." *Ambio* 31 (2002): 102–12.

Rabalais, Nancy N., Nazan Atilla, Claire Normandeau, and R. Eugene Turner. "Ecosystem History of Mississippi River-influenced Continental Shelf Revealed through Preserved Phytoplankton Pigments." *Marine Pollution Bulletin* 49 (2004): 537–47.

Rabalais, Nancy N., R. Eugene Turner, Quay Dortch, William J. Wiseman Jr., and Barun K. Sen Gupta. "Nutrient Changes in the Mississippi River and System Responses on the Adjacent Continental Shelf." *Estuaries* 19 (1996): 386–407.

Rabalais, Nancy N., R. Eugene Turner, and Donald Scavia. "Beyond Science into Policy: Gulf of Mexico Hypoxia and the Mississippi River." *BioScience* 52 (2002): 129–42.

Sen Gupta, Barun K., R. Eugene Turner, and Nancy N. Rabalais. "Seasonal Oxygen Depletion in Continental-shelf Waters of Louisiana: Historical Record of Benthic Forminifers." *Geology* 24 (1996): pp 227–2

Turner, R. Eugene, and Nancy N. Rabalais. "Changes in Mississippi River Water Quality This Century: Implications for Coastal Food Webs." *BioScience* 41 (1991): 140–47.

———. "Coastal Eutrophication near the Mississippi River Delta." *Nature* 368 (1994): 619–21.

———. "Linking Landscape and Water Quality in the Mississippi River Basin for 200 Years." *BioScience* 53 (2003): 563–72.

Turner, R. E., N. N. Rabalais, and D. Justic. "Predicting Summer Hypoxia in the Northern Gulf of Mexico: Riverine N, P, and Si Loading." *Marine Pollution Bulletin* 52 (2006): 132–48.

Turner, R. Eugene, Nancy N. Rabalais, Dubravko Justic, and Quay Dortch. "Future Aquatic Nutrient Limitations." *Marine Pollution Bulletin* 46 (2003): 1032–34.

———. "Global Patterns of Dissolved N, P and Si in Large Rivers." *Boiogeochemistry* 64 (2003): 297–317.

Papers on ocean pH

Caldeira, Ken, and Michael E. Wickett. "Anthropogenic Carbon and Ocean pH." *Nature* 425 (2003): 365.

———. "Ocean Model Predictions of Chemistry Changes from Carbon Dioxide Emissions to the Atmosphere and Ocean." *Journal of Geophysical Research* 110 (2005): 12 pp.

Doney, S. C., K. Lindsay, K. Caldeira, et al. "Evaluating Global Ocean Carbon Models: The Importance of Realistic Physics." *Global Biogeochemical Cycles* 18 (2004): 22 pp.

Feely, Richard A., Christopher L. Sabine, Kitack Lee, et al. "Impact of Anthropogenic CO_2 on the $CaCO_3$ System in the Oceans." *Science* 305 (2004): 362–66/

Gattuso, J.-P., M. Frankignoulle, I. Bourge, S. Romaine, and R. W. Buddemeier. "Effect of Calcium Carbonate Saturation of Seawater on Coral Calcification." *Global and Planetary Change* 18 (1998): 37–46.

Hughes, T. P., A. H. Baird, J. M. Louh, et al. "Climate Change, Human Impacts, and the Resilience of Coral Reefs." *Science* 301 (2003): 929–33.

Kleypas, J. A., R. W. Buddemeier, D. Archer, J.-P. Gattuso, C. Langdon, and B. N.

Opdyke. "Geochemical Consequences of Increased Atmospheric CO_2 on Coral Reefs." *Science* 284 (1999): 118–20.

Kleypas, J. A., R. W. Buddemeier, and J.-P. Gattuso. "The Future of Coral Reefs in an Age of Global Change." *International Journal of Earth Sciences* 90 (2001): 426–37.

Kleypas, J. A., R.A. Feeley, V. J. Fabry, C. Langdon, C. L. Sabine, and L. L. Robbins. "Impacts of Ocean Acidification on Coral Reefs and Other Marine Calcifiers: A Guide for Future Research." Report of a workshop, sponsored by NSF, NOAA, and the U.S. Geological Survey, held April 18–20, 2005, St. Petersburg, FL. 88 pp.

Kleypas, J. A., and C. Langdon. "Overview of CO_2-induced Changes in Seawater Chemistry." *Proceedings of the 9th International Coral Reef Symposium, Bali, Indonesia, 23–27 Oct. 2000* 2 (2000): 1087–89.

Kolbert, Elizabeth. "The Darkening Sea: What Carbon Emissions Are Doing to the Ocean." *New Yorker*, November 20, 2006, 65–75.

Kuffner, Ilsa B., Andreas J. Andersson, Paul L. Jokiel, Ku'Ulei S. Rodgers, and Fred T. Mackenzie. "Decreased Abundance of Crustose Coralline Algae Due to Ocean Acidification." *Nature Geoscience*, advance online publication (2007), doi:10.1038/ngeo100.

Langdon, C., T. Takahashi, C. Sweeney, et al. "Effect of Calcium Carbonate Saturation State on the Calcification Rate of an Experimental Coral Reef." *Global Biogeochemical Cycles* 14 (2000): 639–54.

Orr, James C., Victoria J. Fabry, Olivier Aumont, et al. "Anthropogenic Ocean Acidification over the Twenty-first Century and Its Impact on Calcifying Organisms." *Nature* 437 (2005): 681–86.

Phinney , J. T., O. Hoegh-Guldberg, J. Kleypas, W. Skirving, and A. Strong. *Coral Reefs and Climate Change: Science and Management*. AGU Monograph Series, Coastal and Estuarine Studies, vol. 61. Washington, DC: American Geophysical Union, 2006.

Royal Society. *Ocean Acidification due to Increasing Atmospheric Carbon Dioxide Policy Document 12/05*. London, UK: Royal Society, 2005.

Sarmiento, J. L., R. Slater, R. Barber, et al. "Response of Ocean Ecosystems to Climate Warming." *Global Biogeochemical Cycles* 18 (2004): 23 pp.

Scheibner, C., and R. P. Speijer. "Decline of Coral Reefs during Late Paleocene to Early Eocene Global Warming." *eEarth Discuss.* 2 (2007): 133–50. http://www.electronic-earth-discuss.net/2/133/2007/

Papers on plankton

Attrill, Martin J., Jade Wright, and Martin Edwards. "Climate-related Increases in Jellyfish Frequency Suggest a More Gelatinous Future for the North Sea." *Limnology and Oceanography* 52 (2007): 480–85.

Beaugrand, Gregory, Philip C. Reid, Frederic Ibanez, J. Alistair Lindley, and Martin Edwards. "Reorganization of North Atlantic Marine Copepod Biodiversity and Climate." *Science* 296 (2002): 1692–94.

Burkill, P. H. (guest editor). "Arabesque: UK JGOFS Process Studies in the Arabian Sea." *Topical Studies in Oceanography: Deep Sea Research Part II* 46 (1999).

Chisholm, Sallie W., and Francois M. M. Morel, eds. "What Controls Phytoplankton

Production in Nutrient-rich Areas of the Open Sea?" *Limnology and Oceanography* 36 (1991): 9 pp.

Edwards, M., D. G. Johns, S. C. Leterme, E. Svendsen, and A. J. Richardson. "Regional Climate Change and Harmful Algal Blooms in the Northeast Atlantic." *Limnology and Oceanography* 51 (2006): 820–29.

Edwards, M., D. G. Johns, P. Licandro, A. W. G. John, and D. P. Stevens. "Ecological Status 2004/2005: Results from the North Atlantic CPR Survey." *SAHFOS Technical Report* 3 (2004): 1–8.

Edwards, Martin, and Anthony J. Richardson. "Impact of Climate Change on Marine Pelagic Phenology and Trophic Mismatch." *Nature* 430 (2004): 881–84.

Falkowski, Paul, and Andrew Knoll, eds. *Evolution of Primary Producers in the Sea.* Toronto: Academic Press, 2007.

Frederiksen, Morten, Martin Edwards, Anthony J. Richardson, Nicholas C. Halliday, and Sarah Wanless. "From Plankton to Top Predators: Bottom-up Control of a Marine Food Web across Four Trophic Levels." *Journal of Animal Ecology* (2006): 10 pp.

Genner, Martin J., David W. Sims, Victoria J. Wearmouth, et al. "Regional Climatic Warming Drives Long-term Community Changes of British Marine Fish." *Proceedings of the Royal Society of London B* 271 (2004): 655–61.

Hawkins, Stephen J., Alan J. Southward, and Martin J. Genner. "Detection of Environmental Change in a Marine Ecosystem—Evidence from the Western English Channel." *Science of the Total Environment* 310 (2003): 245–56.

Herbert, R. J. H., S. J. Hawkins, M. Sheader, and A. J. Southward. "Range Extension and Reproduction of the Barnacle *Balanus perforatus* in the Eastern English Channel." *Journal of the Marine Biological Association of the United Kingdom* 83 (2003): 73–82.

Iglesias-Rodriguez, M. Debora, Christopher W. Brown, Scott C. Doney, et al. "Representing Key Phytoplankton Functional Groups in Ocean Carbon Cycle Models: Coccolithophorids." *Global Biogeochemical Cycles* 16 (2002): 1100.

Lopez-Urrutia, Angel, Elena San Martin, Roger P. Harris, and Xabier Irigoien. "Scaling the Metabolic Balance of the Oceans." *PNAS* 103 (2006): 8739–44.

Marine Climate Change Impacts Partnership (UK). "Annual Report Card 2007." http://www.mccip.org.uk/arc (accessed May 25, 2008).

Owens, N. J. P., C. S. Law, R. F. C. Mantoura, P. H. Burkill, and C. A. Llewellyn. "Methane Flux to the Atmosphere from the Arabian Sea." *Nature* 354 (1991): 293–96.

Pitois, S., and C. J. Fox. "Detecting Changes in Long-term Zooplankton Abundance and Size Structure in the North Atlantic." *ICES Journal of Marine Science* 63 (2006): 785–98.

Portner, Hans O., and Rainer Knust. "Climate Change Affects Marine Fishes through the Oxygen Limitation of Thermal Tolerance." *Science* 315 (2007): 95–97.

Raitsos, Dionysios E., Philip C. Reid, Samantha Lavender, Martin Edwards, and Anthony J. Richardson. "Extending the SeaWiFS Chlorophyll Data Set Back 50 Years in the Northeast Atlantic." *Geophysical Research Letters* 32 (2005): 4 pp.

Reid, P. C., J. M. Colebrook, J. B. L. Matthews, and J. Aiken. "The Continuous

Plankton Recorder: Concepts and History, from Plankton Indicator to Undulating Recorders." *Progress in Oceanography* 58 (2003): 117–73.
Richardson, A. J., and A. W. Walne. "Using Continuous Plankton Recorder Data." *Progress in Oceanography* 68 (2006): 27–74.
Ridgwell, A., J. C. Hargreaves, N. R. Edwards, et al. "Marine Geochemical Data Assimilation in an Efficient Earth System Model of Global Biogeochemical Cycling." *Biogeosciences* 4 (2007): 87–104.
Ridgwell, Andy J., Mark A. Maslin, and Andrew J. Watson. "Reduced Effectiveness of Terrestrial Carbon Sequestration Due to an Antagonistic Response of Ocean Productivity." *Geophysical Research Letters* 29 (2002): 19 pp.
Rusch, Douglas B., Aaron L. Halpern, Granger Sutton, et al. "The Sorcerer II Global Ocean Sampling Expedition: Northwest Atlantic through Eastern Tropical Pacific." *PLoS Biology* 5 (2007): 34.
Rutherford, Scott, Steven D'Hondt, and Warren Prell. "Environmental Controls on the Geographic Distribution of Zooplankton Diversity." *Nature* 400 (1999): 749–53.
Southward, Alan J., O. Langmead, N. J. Hardman-Mountford, et al. "Long-term Oceanographic and Ecological Research in the Western English Channel." *Advances in Marine Biology* 47 (2005): 1–105.
Springer, A. M., J. A. Estes, G. B. van Vliet, et al. "Sequential Megafaunal Collapse in the North Pacific Ocean: An Ongoing Legacy of Industrial Whaling?" *PNAS* 100 (2003): 12223–28.

Papers on fish

Berkes, F., T. P. Hughes, R. S. Steneck, et al. "Globalization, Roving Bandits, and Marine Resources." *Science* 311 (2006): 1557–58.
Boddeke, R., and B. Vingerhoed. "The Anchovy Returns to the Wadden Sea." *ICES Journal of Marine Science* 53 (1996): 1003–7.
Devine, Jennifer A., Krista D. Baker, and Richard L. Haedrich. "Fisheries: Deep-sea Qualify as Endangered." *Nature* 439 (2006).
Hoffmann, R. C. "A Brief History of Aquatic Resource Use in Medieval Europe." *Helgoland Marine Research* 59 (2005): 22–30.
Hutchings, Jeffrey A. "Collapse and Recovery of Marine Fishes." *Nature* 406 (2000): 882–85.
———. "Spatial and Temporal Variation in the Density of Northern Cod and a Review of Hypotheses for the Stock's Collapse." *Canadian Journal of Fisheries and Aquatic Sciences* 53 (1996): 943–62.
Hutchings, Jeffrey A., and Ransom A. Myers. "What Can Be Learned from the Collapse of a Renewable Resource? Atlantic Cod, *Gadus morhua*, of Newfoundland and Labrador." *Canadian Journal of Fisheries and Aquatic Sciences* 51 (1994): 2126–46.
James Ford Bell Library. "Olaus Magnus Map of Scandinavia." http://bell.lib.umn.edu/olaus/ (accessed May 26, 2008).
Lotze, Heike K. "Repetitive History of Resource Depletion and Mismanagement: The Need for a Shift in Perspective." *Marine Ecology Progress Series* 274 (2004): 282–85.

———. "Rise and Fall of Fishing and Marine Resource Use in the Wadden Sea, Southern North Sea." *Fisheries Research* 87 (2007): 208–18.

Lotze, Heike K., Hunter S. Lenihan, Bruce J. Bourque, et al. "Depletion, Degradation, and Recovery Potential of Estuaries and Coastal Seas." *Science* 312 (2006): 1806–9.

Lotze, Heike, and Inka Milewski. "Two Centuries of Multiple Human Impacts and Successive Changes in a North Atlantic Food Web." *Ecological Applications* 14 (2004): 1428–47.

Lotze, Heike, and Karsten Reise, eds. *Helgoland Marine Research*. Vol. 59, no. 1. Bremerhaven, Germany: Springer-Verlag Berlin Heidelberg and Alfred Wegener Institute for Polar and Marine Research, April 2005.

Myers, Ransom A., and Boris Worm. "Rapid Worldwide Depletion of Predatory Fish Communities." *Nature* 423 (2003): 280–83.

Pauly, Daniel, Jackie Alder, Elena Bennett, Villy Christensen, Peter Tyedmers, and Reg Watson. "The Future for Fisheries." *Science* 302 (2003): 1359–61.

Pauly, Daniel, and Maria-Lourdes Palomares. "Fishing Down Marine Food Web: It Is Far More Pervasive Than We Thought." *Bulletin of Marine Science* 76 (2005): 197–212.

Pauly, Daniel, and Reg Watson. "Counting the Last Fish." *Scientific American* 289 (2003): 42–47.

Pauly, Daniel. "Major Trends in Small-scale Marine Fisheries, with Emphasis on Developing Countries, and Some Implications for the Social Sciences." *MAST* 4 (2006): 7–22.

Rosenberg, Andrew A., W. Jeffrey Bolster, Karen E. Alexander, William B. Leavenworth, Andrew B. Cooper, and Matthew G. McKenzie. "The History of Ocean Resources: Modeling Cod Biomass Using Historical Records." *Frontiers in Ecology and the Environment* 3 (2005): 84–90.

Schrope, Mark. "Oceanography: The Real Sea Change." *Nature* 443 (2006): 622–24.

State of World Fisheries and Aquaculture, 2006. Rome: United Nations Food and Agriculture Organization, 2006.

Sumaila, U. R., and D. Pauly, eds. *Catching More Bait: A Bottom-up Re-estimation of Global Fisheries Subsidies*. Fisheries Centre Research Reports 14, no. 6 (2006), 2nd version. Vancouver, BC: University of British Columbia.

Ward, Peter, and Ransom A. Myers. "Shifts in Open-ocean Fish Communities Coinciding with the Commencement of Commercial Fishing." *Ecology* 86 (2005): 835–47.

Worm, B., E. B. Barbier, N. Beaumont, et al. "Impacts of Biodiversity Loss on Ocean Ecosystem Services." *Science* 314 (2006): 787–90.

Worm, Boris, Heike K. Lotze, and Ransom A. Myers. "Predator Diversity Hotspots in the Blue Ocean." *PNAS* 100 (2003): 9884–88.

Worm, Boris, Marcel Sandow, Andreas Oschlies, Heike Lotze, and Ransom A. Myers. "Global Patterns of Predator Diversity in the Open Oceans." *Science* 309 (2005): 1365–69.

Zeller, Dirk, and Daniel Pauly. "Good News, Bad News: Global Fisheries Discards Are Declining, but so Are Total Catches." *Fish and Fisheries* 6 (2005): 156–59.

Papers on ocean and climate change

Berman, J. Michael, Kevin R. Arrigo, and Pamela A. Matson. “Agricultural Runoff Fuels Large Phytoplankton Blooms in Vulnerable Areas of the Ocean.” *Nature* 434 (2005): 211–14.

Boyer, Timothy B., S. Levitus, J. I. Antonov, R. A. Locarnini, and H. E. Garcia. “Linear Trends in Salinity for the World Ocean 1995–1998.” *Geophysical Research Letters* 32 (2005): 4 pp.

Boyle, Edward A. “Characteristics of the Deep Ocean Carbon System during the Past 150,000 Years.” *PNAS* 94 (1997): 8300–8307.

Buermann, Wolfgang, Benjamin R. Lintner, Charles D. Koven, et al. “The Changing Carbon Cycle at Mauna Loa Observatory.” *PNAS* 104 (2007): 4249–54.

Canadell, Josep G., Corinne Le Quéré, Michael R. Raupach, et al. “Contributions to Accelerating Atmospheric CO_2 Growth from Economic Activity, Carbon Intensity, and Efficiency of Natural Sinks.” *PNAS* 104 (2007): 18866–70.

Corredor, Jorge E., and Julio M. Morell. “Seasonal Variation of Physical and Biogeochemical Features in Eastern Caribbean Surface Water.” *Journal of Geophysical Research* 106 (2001): 4517–25.

Curry, Ruth, Bob Dickson, and Igor Yashayaev. “A Change in the Freshwater Balance of the Atlantic Ocean over the Past Four Decades.” *Nature* 426 (2003): 826–29.

Dickson, Bob, Igor Yashayaev, Jens Meincke, Bill Turrell, Stephen Dye, and Juergen Holfort. “Rapid Freshening of the Deep North Atlantic Ocean over the Past Four Decades.” *Nature* 416 (2002): 832–37.

Friedlingstein, Pierre, and Susan Solomon. “Contributions of Past and Present Human Generations to Committed Warming Caused by Carbon Dioxide.” *PNAS* 102 (2005): 10832–36.

Galloway, James N. “The Nitrogen Cascade.” *BioScience* 53 (2003): 341–56.

Goreau, Thomas J., Raymond L. Hayes, and Don McAllister. “Regional Patterns of Sea Surface Temperature Rise: Implications for Global Ocean Circulation Change and the Future of Coral Reefs and Fisheries.” *World Resource Review* 17 (2005): 350–73.

Govindasamy, B., S. Thompson, A. Mirin, M. Wickett, K. Caldeira, and C. Delire. “Increase of Carbon Cycle Feedback with Climate Sensitivity: Results from a Coupled Climate and Carbon Cycle Model.” *Tellus* 57B (2005): 153–63.

Halpern, Benjamin S., Shaun Walbridge, Kimberly A. Selkoe, et al. “A Global Map of Human Impact on Marine Ecosystems.” *Science* 319 (2008): 948–52.

Mayhew, Peter J., Gareth B. Jenkins, and Timothy G. Benton. “A Long-term Association between Global Temperature and Biodiversity, Origination and Extinction in the Fossil Record.” *Proceedings of the Royal Society of London B* 275 (2008): 47–54.

McLaughlin, John F., Jessica J. Hellmann, Carol L. Boggs, and Paul R. Ehrlich. “Climate Change Hastens Population Extinctions.” *PNAS* 99 (2002): 6070–74.

Pahlow, Markus, and Ulf Riebesell. “Temporal Trends in Deep Ocean Redfield Ratios.” *Science* 287 (2000): 831–33.

Raupach, Michael, Gregg Marland, Philippe Ciais, et al. “Global and Regional Drivers of Accelerating CO_2 Emissions.” PNAS 104 (2007): 10288–93.

Scholze, Marko, Wolfgang Knorr, Nigel W. Arnell, and I. Colin Prentice. "A Climate-change Risk Analysis for World Ecosystems." *PNAS* 103 (2006): 13116–20.

Thompson, Richard C. "Lost at Sea: Where Is All the Plastic?" *Science* 304 (2004): 838.

Williams, John W., Stephen T. Jackson, and John E. Kutzbach. "Projected Distributions of Novel and Disappearing Climates by 2100 AD." *PNAS* 104 (2007): 5738–42.

Papers on the Paleocene/Eocene Thermal Maximum

Brinkhuis, Henk, Stefan Schouten, Jonathan Bujak, et al. "Episodic Fresh Surface Waters in the Eocene Arctic Ocean." *Nature* 441 (2006): 606–9.

Huber, Matthew, and Rodrigo Caballero. "Eocene El Nino: Evidence for Robust Tropical Dynamics in the 'Hothouse.'" *Science* 299 (2003): 877–81.

Jaramillo, Carlos, Milton J. Rueda, and German Mora. "Cenozoic Plant Diversity in the Neotropics." *Science* 311 (2006): 1893–96.

Kerr, Richard A. "Global Climate Change: The Atlantic Conveyor May Have Slowed, but Don't Panic Yet." *Science* 310 (2005): 1403–4.

Najjar, Raymond G., Giang T. Nong, Dan Seidov, and William H. Peterson. "Modeling Geographic Impacts on Early Eocene Ocean Temperature." *Geophysical Research Letters* 29 (2002): 1750.

Pagani, Mark, James C. Zachos, Katherine H. Freeman, Brett Tipple, and Stephen Bohaty. "Marked Decline in Atmospheric Carbon Dioxide Concentrations during the Paleogene." *Science* 309 (205): 600–603.

Pagani, Mark, Ken Caldeira, David Archer, and James C. Zachos. "An Ancient Carbon Mystery." *Science* 314 (2006): 1556–57.

Pagani, Mark, Michael A. Arthur, and Katherine H. Freeman. "Miocene Evolution of Atmospheric Carbon Dioxide." *Paleoceanography* 14 (1999): 273–92.

Pagani, Mark, Michael A. Arthur, and Katherine H. Freeman. "Variations in Miocene Phytoplankton Growth Rates in the Southwest Atlantic: Evidence for Changes in Ocean Circulation." *Paleoceanography* 15 (2000): 486–96.

Pagani, Mark. "Arctic Hydrology during Global Warming at the Palaeocene/Eocene Thermal Maximum." *Nature* 442 (2006): 671–75.

Pagani, Mark. "The Alkenone-CO_2 Proxy and Ancient Atmospheric Carbon Dioxide." *Philosophical Transactions of the Royal Society London* A 360 (2002): 609–32.

Schmidt, Gavin A., and Drew T. Shindell. "Atmospheric Composition, Radiative Forcing, and Climate Change as a Consequence of a Massive Methane Release from Gas Hydrates." *Paleoceanography* 18 (2003): 9 pp.

Schwartz, Peter, and Doug Randall. "An Abrupt Climate Change Scenario and Its Implications for United States National Security, October 2003." GBN Global Business Network. http://gbn.com (accessed May 27, 2008).

Sluijs, Appy, Stefan Schouten, Jens Matthiessen, et al. "Subtropical Arctic Ocean Temperatures during the Palaeocene/Eocene Thermal Maximum." *Nature* 441 (2006): 610–13.

Sriver, Ryan L., and Matthew Huber. "Observational Evidence for an Ocean Heat Pump Induced by Tropical Cyclones." *Nature* 447 (2007): 577–80.

Stoll, Heather. "Climate Change: The Arctic Tells Its Story." *Nature* 441 (2006): 579–81.
Witze, Alexandra. "Bad Weather Ahead." *Nature* 441 (2006): 564–66.
Zachos, James, Mark Pagani, Lisa Sloan, Ellen Thomas, and Katharina Billups. "Trends, Rhythms, and Aberrations in Global Climate 65 MA to Present." *Science* 292 (2001): 686–93.

Papers on China

CCICED, Secretariat. "President Hu Jintao Expounds China's Stance on Climate Change at APEC Meeting." *CCICED Update* 23 (2007): 5 pp.
China's National Climate Change Programme. Beijing: National Development and Reform Commissions, 2007.
Hanson, Arthur J., and Claude Martin. *One Lifeboat: China and the World's Environment and Development*. Winnipeg, MB: International Institute for Sustainable Development, 2006.
Martinot, Eric, and Jungfeng Li. *Powering China's Development: The Role of Renewable Energy*. Washington, DC: Worldwatch Institute, 2007.
National Development and Reform Commission of China. "National Climate Change Program." http://www.china.org.cn/english/environment/213624.htm, June 4, 2007.
Saevarsson, Haflioi. *China Seafood Industry Report*. Shanghai: Glitnir Research Division, 2007.
State Council Information Office of China. "White Paper on Energy." http://www.china.org.cn/english/environment/236955.htm, December 26, 2007.
Wang, Hongying. "China's Changing Approach to Sustainable Development." *Development* 50 (2007): 36–43.
World Energy Outlook 2007. Paris: International Energy Agency, 2007.

Papers on Zanzibar

Cohen, Ilana. *Mollusc Collecting and Bivalve Farming: Economics and Strategies in Two Zanzibari Villages*. Stone Town, Zanzibar, Tanzania: School for International Training, 2005.
Jiddawi, Narriman S., and Marcus C. Ohman. "Marine Fisheries in Tanzania." *AMBIO: A Journal of the Human Environment* 31 (2002): 518–27.
Kariger, Patricia K., Rebecca J. Stoltzfus, Deanna Olney, et al. "Iron Deficiency and Physical Growth Predict Attainment of Walking but Not Crawling in Poorly Nourished Zanzibari Infants." *Journal of Nutrition* 135 (2005): 814–19.
Richmond, M. D., and J. Francis. *Marine Science Development in Tanzania and Eastern Africa: Proceedings of the 20th Anniversary Conference on Advances in Marine Science in Tanzania, 28 June–1 July 1999, Zanzibar, Tanzania*. Stone Town, Zanzibar: Institute of Marine Sciences, University of Dar es Salaam, and Western Indian Ocean Marine Science Association, 2001.
Tanzania State of the Coast 2001: People and the Environment. Tanzania Coastal

Management Partnership Science and Technical Working Group, Working Document 5059. Dar es Salaam, 2001.

Van Der Elst, Rudy, Bernadine Everett, Narriman Jiddawi, Gerald Mwatha, Paula Santana Afonso, and David Boulle. "Fish, Fisher and Fisheries of the Western Indian Ocean: Their Diversity and Status: A Preliminary Assessment." *Philosophical Transactions of the Royal Society A* 363 (2005): 263–84.

Index